Detection of Explosives and Landmines

NATO Science Series

A Series presenting the results of scientific meetings supported under the NATO Science Programme.

The Series is published by IOS Press, Amsterdam, and Kluwer Academic Publishers in conjunction with the NATO Scientific Affairs Division

Sub-Series

I. Life and Behavioural Sciences	IOS Press
II. Mathematics, Physics and Chemistry	Kluwer Academic Publishers
III. Computer and Systems Science	IOS Press
IV. Earth and Environmental Sciences	Kluwer Academic Publishers
V. Science and Technology Policy	IOS Press

The NATO Science Series continues the series of books published formerly as the NATO ASI Series.

The NATO Science Programme offers support for collaboration in civil science between scientists of countries of the Euro-Atlantic Partnership Council. The types of scientific meeting generally supported are "Advanced Study Institutes" and "Advanced Research Workshops", although other types of meeting are supported from time to time. The NATO Science Series collects together the results of these meetings. The meetings are co-organized bij scientists from NATO countries and scientists from NATO's Partner countries – countries of the CIS and Central and Eastern Europe.

Advanced Study Institutes are high-level tutorial courses offering in-depth study of latest advances in a field.
Advanced Research Workshops are expert meetings aimed at critical assessment of a field, and identification of directions for future action.

As a consequence of the restructuring of the NATO Science Programme in 1999, the NATO Science Series has been re-organised and there are currently Five Sub-series as noted above. Please consult the following web sites for information on previous volumes published in the Series, as well as details of earlier Sub-series.

http://www.nato.int/science
http://www.wkap.nl
http://www.iospress.nl
http://www.wtv-books.de/nato-pco.htm

Series II: Mathematics, Physics and Chemistry – Vol. 66

Detection of Explosives and Landmines

Methods and Field Experience

edited by

Hiltmar Schubert

Fraunhofer-Institut für Chemische Technologie (ICT),
Pfinztal, Germany

and

Andrey Kuznetsov

V.G. Khlopin Radium Institute,
St. Petersburg, Russia

Kluwer Academic Publishers

Dordrecht / Boston / London

Published in cooperation with NATO Scientific Affairs Division

Proceedings of the NATO Advanced Research Workshop on
Detection of Explosives and Landmines
St. Petersburg, Russia
9–14 September 2001

A C.I.P. Catalogue record for this book is available from the Library of Congress.

ISBN 1-4020-0692-6

Published by Kluwer Academic Publishers,
P.O. Box 17, 3300 AA Dordrecht, The Netherlands.

Sold and distributed in North, Central and South America
by Kluwer Academic Publishers,
101 Philip Drive, Norwell, MA 02061, U.S.A.

In all other countries, sold and distributed
by Kluwer Academic Publishers,
P.O. Box 322, 3300 AH Dordrecht, The Netherlands.

Printed on acid-free paper

Printed in the Netherlands.

Table of Contents

Preface

This ARW is the third NATO-sponsored workshop on Explosives Detection and Humanitarian Demining. The previous events were

- Detection and Destruction of Anti-Personnel Landmines
  Moscow, 1997
- Explosives Detection and Decontamination of the Environment
  Prague, 1997.

Over the last decade applied research in Humanitarian Demining has made progress to some extend, but according to the tremendous tasks of Demining and the lack of scientific methods for practical detection of explosive devices, research activities are still of the same importance than ever before. Concerning countermeasures against terrorism the detection of explosives is one of the keyfactors, but the practical applications are not sufficient solved. An international exchange of research results are therefore urgent, to find out the most promising measures for application. The coincidence of this ARW and the terrible disaster of New York and Washington may demonstrate the importance of this task.

In consequence the explosive device detection technologies can make a major contribution to collective, family and individual security. In developed countries, these technologies provide a strong deterrent and preventative measure against terrorist threats. In less developed regions, they can improve individual, institutional and state security, lessening the insecurity that motivates many terrorists acts. The elimination of landmine threats is just one of many ways of achieving this. However our attempts to meet the extremely difficult technical challenges posed by landmine and UXO contamination are inevitably leading us to new technological approaches.

If these new approaches are pursued vigorously they will definitely enhance our capabilities to detect hidden weapons, explosive devices, even weapons of mass destruction.

Recent events have underlined the pressing need for improved security measures.

Technological innovation is continually opening new avenues for terrorist acts. No countermeasures can be perfect. Therefore security ultimately relies on maintaining greater knowledge and technological capacity than terrorist groups.

There is a continuing need for research to extend our presently reduced lead and restore an acceptable technological edge.

An effective deterrent to further use of landmines and to guard against terrorism requires three elements: technological capability, credibility that it will be deployed, and communicating this to those whom we wish to deter. However, we do not have a fully effective technological capability.

The problem of reliable detection of explosives, whether in buried landmines, or hidden in cars, aircraft cargo or baggage, or carried by suicide bombers, has not yet been solved. Resources provided for research until now have not been sufficient to meet he challenge. While the technical problems pose a great challenge to our ingenuity, they are not insurmountable. More resources are needed.

Further, continuing research activities are needed to maintain the expertise and technology capability to respond quickly to new and unforeseen threats.

The goal of the workshop was to bring Eastern and Western scientists with experience in landmine and explosive detection. The main task was to review the state of the art in landmine and explosive detection technologies, and to consider whether there are alternatives methods or technologies that should be investigated.

Papers describing physical, chemical, and bio-chemical methods were presented, covering the complex aspects of mine detection, e.g. shape, materials and explosives.

The papers covered a wide range of different detection methods, and included sensor-fusion experiments such as MMSR and HOPE.

Discussions focused on the needs expressed by mine action agencies and of the technical realization of the methods. It seems likely that some existing methods could be effective for particular tasks in specific environmental conditions. However, current sensor technologies are not sufficient for most mine action tasks. However, no single sensor system will be able to fulfilling all the necessary needs. Therefore, methods for using multiple sensor approaches are essential, for example, sensor fusion.

The major needs from mine action agencies that were discussed include:

1. Technologies for mapping contaminated areas and assessing the effectiveness of remediation efforts,
2. Quality assurance sensor for mine clearance work, and
3. Detection and location of individual mines.

Some of the significant conclusions and results of this meeting were:

a) There is no known method that will the requirement to detect minimum metal mines at a depth of 200mm in all conditions. However intermediate solutions that fall short of this requirement may still meet certain niche needs of the mine action and the military peace-keeping or peace-enforcement communities. For example, a partially effective technology road clearance, assisted by mine detection dogs, could effectively deter mine-laying activities in favour of other strategies that do not pose the same problems as mines in the post-conflict recovery phase. A ground penetrating radar sensor could be helpful in locating large anti-vehicle mines detected by dogs: the dogs cannot accurately locate most mines.
b) Up to this moment, no single sensor seems to fulfil the need to detect individual mines in all possible scenarios. This means a multi-sensor system is needed, in which the outcomes of several sensors are combined together. Furthermore, the influence of the environment requires different sensors to be used in different scenarios. All these remarks together suggest that there is a need for an open architecture multi-sensor (data fusion) system: this can be described as a "tool box" approach.
c) Scientists should not work without a clear understanding of the problems to be solved. Therefore it is proposed that scientists must develop strong links with the end-users and visit real mine fields, and that funding should be provided specifically to assist this.
d) To evaluate the potential for different methods, there is a need for standardized simulants, test fields, and test procedures. This could be realized in connection with the ITEP initiative.

e) There are technologies that could help solve humanitarian demining problems but the commercial potential is limited and the realization of these systems is taking too much time. Present government policies that favour research on commercially attractive technologies need to be adjusted to overcome this difficulty.

f) Given the likely long term of the problem of mine contamination, we think that related fundamental scientific research could be useful. More research on, for example, olfactory sensing in primates, electromagnetic properties of soils and diffusion of explosive molecules and particles in the environment, for example, could be very helpful.

g) The workshop has helped co-operation and cross-fertilization between the research institutes of different countries (and especially between East and West). Several collaborative research projects were discussed and initiated at the workshop.

Co-chairmen:

Hiltmar Schubert
Andrey Kuznetsov

Acknowledgement

Dr. Andrey Kuznetsov from the Khlopin Institute in St. Petersburg – my co-chairman – has had the idea for this ARW and the international demining community should thank him for this initiative, for the local organization of this event and for the collection of all the eastern contributions to this research area. Thanks also to the director of the Khlopin Institute, Prof. Dr. Rimski-Korsakov, for his support.

I thank my western colleagues for following my invitation to present and discuss their new research results in St. Petersburg.

I am very obliged to Prof. Trevelyan, Dr. Verlinde and Dr. Krausa for their support not only in the formulation of the conclusions and results of the workshop, but also in the needs for future activities, which were laid down in the preface of this book.

Last but not least I thank the Advisory Panel on Security-Related Civil Science & Technology and its Programme Director for the support and interest in this ARW.

Hiltmar Schubert

Chemical methods for the detection of mines and explosives

M. Krausa, H. Massong, P. Rabenecker and H. Ziegler

Fraunhofer-Institut für Chemische Technologie (ICT)

Joseph-von-Fraunhofer-Straße 7, 76327 Pfinztal, Germany

Abstract

Up to now landmine detection and clearance is a time consuming and dangerous process. The most time consuming step is the detection and location of the mines. One of the most successful mine and explosive detection systems are the so called 'sniffer' dogs. It is known that noses of dogs are much more sensitive compared to the noses of human beings. Therefore dogs are able to find mines despite the low concentrations of the vaporizing explosives. Until today there is a controversial discussion if dogs are able to smell the pure explosives or the odor bouquet of the mines.

In comparison with other methods the detection of explosives by dogs is a fast process. On the other hand dogs are living beings and their behavior and skills are influenced by numerous parameters (e.g. age, sex, trainer, climate, daily condition, etc.). To overcome these problems concerning the detection of explosives by dogs, chemical sensor systems are planned and tested which are sensitive enough to smell mines using different chemical techniques. Because of the low vapor pressures of the common explosives and the complex composition of the samples the chemical smelling is a demanding scientific problem.

The presented paper describes the requirements of chemical sensor systems for the detection of explosives and mines on the basis of the results received by the investigation of sniffer dogs and on information of the concentration and composition of the gas phase over mines by on-site measurements. A brief overview of chemical, biochemical and electrochemical methods for the trace detection of explosives and mines will be given.

H. Schubert and A. Kuznetsov (eds.), Detection of Explosives and Landmines, 1–19.

Introduction

It is estimated that more than 110 million active mines are scattered in 70 countries around the world and that for every mine cleared 20 are laid [1]. In 1994, approximately 100.000 mines were removed, while approx. 2 million were laid. On the basis of the today clearing rate the landmine problem will be a problem for long years. Therefore the necessity to enhance the detection and clearing is obvious.

Presently mines are detected individually by prodding, metal detectors or sniffer dogs. In comparison to sniffer dogs prodding is a slow and dangerous event. Metal detectors works well with metal-cased mines, but the metal content in modern mines has been increasingly replaced by plastics. Therefore modern mines will be undetectable by metal detectors. Moreover the detection rate using metal detectors is reduced because of the high number of false alarms as a result of metal fragments buried in soils. For one detected mine the false alarm rate varies between 100 and 1000. Although metal detectors are very successful the clearance rate could not be enhanced by metal detectors drastically.

In comparison to that dogs are very successful in detecting explosives and narcotics during police missions. Special trained sniffer dogs works very well in the field of landmine detection. Many countries worldwide established special projects for the training and use of sniffer dogs [2]. With the use of the sniffer dogs the false alarm rate could be reduced and the detection rate could be enhanced. On the other hand dogs are living beings and their performance is influenced by various factors as was summarized by Trevelyan et al., e.g. sex, age, training regime, health of dog and dog handler, living conditions of dogs and handler [3].

Obviously dogs detect mines by their smell. Up to now it is a point of discussion if the dogs really smell the explosive vapor or a mine like odor bouquet. Nevertheless smelling mines seemed to be more successful than other methods. Therefore the development of chemical methods, which overcomes the special problems of dogs, seemed to be a very promising way to enhance the detection rate. Today a lot of different detection methods and strategies are used for this purpose, but several factors influences the efficiency of these different methods. In view of the further development of improved methods for artificial smelling mines some general questions should be answered:

How smells a mine?

Which smell is identified as mine like by dogs?

What are the TNT-concentrations and the concentrations of other substances above buried mines? And how do they change with time?

How smells a mine?

Mines differ strongly by their shape and materials [4]. The shape of the mines will presumably not influence the odor bouquet but the materials will do. In modern mines the metals are replaced by different plastics, e.g. polyvinylchloride or polystyrene and additionally different softener. In the case of metal mines different oils are used to protect the mines against corrosion. All these chemicals will have a different smell. In addition the explosives (pure TNT, RDX, Picric acid) and the compositions (TNT and RDX) varies, beside of that different waxes and oils are included also. Most of the buried mines are filled with TNT. Depending on the production processes different contamination are include for example DNT, DNB. Therefore it seemed to be very complicated to describe the smell of mines in an easy and common way.

The most common aspect comparing all the mines with their different compositions seems to be the included TNT. In view of the development of chemical methods most researchers focuses their work on the detection of TNT. The major problem of the TNT-detection from the vapor phase is the very low vapor concentration of TNT. The vapor pressure of TNT (Tab.1) at 20 °C is approx. 7 ppb [5] and the concentrations above mines are much smaller as will be presented later. If it is assumed that the dogs smell mines the question arises what aspect of the mine did dogs smell. Does a dog identify a mine by smelling TNT only or will the dog smell the chemicals with a higher vapor pressure than TNT, e.g. DNT, DNB, waxes, oils? In our opinion the answers to these questions will give important information for the development of new sensors system for the mine detection.

Because of the low vapor pressure the detection of TNT is a high sophisticated process. Moreover if research and development focus on the most difficult to detect substances and the dogs which are more successful, focuses on other chemicals or an odor bouquet research will lose the competition. Therefore a second questions arises:

Which smell is identified as mine like by dogs?

At the Canine Olfactory Laboratory at the Auburn University research is conducted to answer this question [6]. For example they presented a gas chromatogram of a PMA-1A landmine [7]. As can be seen from a published chromatogram no TNT is found in the vapor phase. On the other hand 1.2 DNB, 1.3 DNB, 2.4 DNT, 2.6 DNT and 3.5 DNT were found in the vapor phase. This result impressively demonstrates that a sensor system which is optimized in detecting small TNT-concentrations will not find this mine! In addition this result gives a tip that dogs might not identify mines by smelling TNT alone. Therefore a sensor system which focuses on TNT alone will not be successful in detecting different mines.

Tab. 1 Vapor pressures of relevant chemicals

Substance	ng/L
2,4,6 TNT	70,0
1,3 DNB	8.140,0
2,4 DNT	1.440,0
2,6 DNT	5.560,0
RDX	0,04
HMX	0,38
PETN	0,09

Research at Auburn University is focused on the question what does a dog smell also. Therefore experiments with nitroglycerine smokeless powder were conducted. The dogs were trained with smokeless powder [6]. Afterwards the dogs should identify the gas phase which smells like the powder. The gas phase was composed by different chemicals. The dogs used the combination of acetone, toluene and limonene as the nitroglycerine specific odor. On the basis of this experiment it can be concluded that the dogs will use a odor bouquet to detect mines and it is not clear which substances are delivering the mine like odor. In view of research and development of chemical sensor system this result gives the tip that it would be more successful to estimate the odor bouquet of mines and to combine the most powerful sensor systems for the components of this odor. Moreover most of the chemical methods used today to detect mines are developed in view of TNT but it seems to be a better way that R&D work would be focused on the mine smell rather the pure TNT alone. Maybe this would be a faster way to develop an effective chemical mine detector. Today works are conducted to use several different chemical methods for pattern recognition.

What are the TNT-concentrations above buried mines?

For the development and evaluation of chemical sensors for mine detection the expected TNT-concentrations above the soils in the vapor phase should be taken into account. A first tip will be given by the low vapor pressure of TNT (s. Tab. 1). Moreover we can assume that the expected vapor concentrations above the buried mines are much lower. The US Army Corps of Engineers examined the TNT, DNB and DNT concentrations found on mines surface of different mines [8]. Their results show for most cases concentration of DNB, DNT and TNT in the range of approx. 0.2 – 30 ng/cm^2. The maximum TNT-concentration on top of a TMA-5 and TMM-1 were higher, 374 ng/cm^2 and 89.9 ng/cm^2. Nevertheless more important are information about the vapor concentrations above buried mines at the soil surface. The US Army Corps of Engineers conducted examinations with buried military grade TNT in different soils (sand, silt, clay) and with varying moistures [9]. After 6 days stored at 23 °C the vapor phase was sampled for 1 min above the buried explosive. There results clearly show that the vapor concentrations of the chemicals e.g. DNB, DNT and TNT were strongly influenced by the soil type and the moisture content. In all cases the TNT-concentrations compared with the vapor concentrations of the other chemicals were much lower. The highest concentrations were observed for DNB and DNT in sand. In clay no TNT was detected in the vapor phase. These very interesting test give a hint that in view of mine detection by TNT detection alone seems to be a problematic way. The Draper Laboratory investigated the concentrations of TNT, DNT, DNB, ADNT in soils at different distances from a buried TMA-5 mine [10]. The highest concentrations of all substances were found directly under the mine body. With increasing distance from the mine in direction to the surface the concentrations decreased generally. No TNT or DNB were found at the soil surface but DNT and ADNT. This result is supported by examinations of Kjellström et al. from FOA [11]. They analyzed the vapor and solid phases above real buried mines in Cambodia and Bosnia-Herzegovina. In the gas samples 2.4 DNT and 2.6 DNT were found only. TNT was detected in no vapor sample of mine affected areas. The authors assume that possibly the sensitivity of the used analytical methods is not high enough to find any traces of TNT.

Sniffer dogs locate the landmines not in all cases exactly. These observation was clarified by Sandia National Labs [12]. Obviously there are different mechanisms of TNT-transportation from the mine to the surface. Beside of pure diffusion in a vapor phase also vapor-liquid partitioning, liquid-solid and vapor-solid partitioning influence the transport of TNT to the surface. Because of the different mechanism the highest TNT-concentration at the soil surface must not necessarily be direct above the mine. The highest surface concentration might be displaced up to one meter beside the mine. This effect which influences dogs also should be taken into account developing and using chemical sensors. In addition the transport of TNT from the buried mine up to the surface is time consuming. Dog handler reported that freshly buried mines are not detected by sniffer dogs. It can be assumed that the dogs will find the mines when the chemicals emanating from the mine reached the surface.

Another parameter which has to be taken into account is biotransformation of TNT [13]. Depending on the microclimate, temperature or moisture microorganism transform TNT into other products, e.g. DAT, ADNT etc.. Depending on the microorganism in the soils it seems possible that the odor of a mine will change. This means that mines in different areas may smell different. A hint that may underline this assumption is the observation that dogs trained in one area have to be retrained if they should work in areas with different vegetation, soils and climates.

Most of these findings have a direct influence on the development and expectations of sensor systems for mine detection. Obviously sensor systems should be as sensitive as possible, independent of the substance which should be detected. The vapor concentration of mine typical substances e.g. TNT, DNT or DNB are in the range of some ng/L (ppb) or lower. The qualitative detection and estimation of these low concentration is a high sophisticated goal. On the other hand the results of the different groups demonstrates that the concentration of DNT and DNB are much higher and therefore it seems to be easier to detect them. It is plausible that dogs uses more than one substance to identify a smell as mine like. It can be assumed that it is much more easy for the dogs to notice substances with higher vapor pressures and with higher concentration in the vapor phase like DNT and DNB as in comparison with TNT. Moreover information about the complete smell of mines that includes softener, waxes, oils and other chemicals after our knowledge are not available. Maybe some of these components could be used as the leading substance to detect the odor bouquet of a mine.

Sensor systems

There are a lot of different chemical sensor systems which are planned to be used for mine detection. Most of them, including our electrochemical system, are optimized to detect small concentrations of TNT but for our opinion the detection of DNT or DNB or the fusion of different sensor systems as pattern recognition would be a better way to detect mines, may be as successful as dogs do. A brief description, which covers some of the chemical systems and which is surely incomplete will be given in the following.

Rats

Beside of dogs other animals are tested for the purpose of mine detection. Rats seem to be very promising candidates in this field. In Belgium a project was started to train rats for mine detection [14]. There are several advantages compared to dogs, e.g.: better sense to smell, resistant to tropical disease, independent of handler and cheaper to keep and to train. Moreover because of their small weight and volume multiple animals could be taken to the field. In opposite to dogs the rats are not rewarded by playing but are strongly motivated by food. The rats slightly scratched the surface only and indicate in this way the position of the mine. APOPO suggested two different ways to use rats for mine detection. One is to reduce the area of suspected mine areas and second the direct detection of mines. First tests are obviously conducted in Tanzania. In addition a test an a mine field which consist of hundreds of real landmines buried at different times and depths was conducted with rats. Menschen gegen Minen (MGM) monitored the search and says that astounding 100% of the mines were detected [15]. Rats seem to be a very good tool for the future mine detection. It is claimed by APOPO that it will take two more years until free-running rats could be used for this purpose.

The customer wishes that a technical sensor displaced far away of the TNT-scource should detect the TNT. If we compare this wish with the kind of detection process of dogs and rats we can see that both animals have to place their sensitive noses near the ground. The dogs smell in a distance of approx. 5-10 cm above and sometimes closer to the surface. The distance between the surface and the rats nose can be assumed as nearer than for dogs. Because of that it can be assumed that technical sensors should be in close distance to the surface also, but not in contact, to the surface.

Bees

Another way to use living beings for mine detection was suggested by different research groups in the USA funded by a program of DARPA. There are different projects in the US which focuses on insects for detection of traces of chemicals. At the Pacific Northwest National Laboratory and at the University of Montana research is conducted to use bees for mine detection [16]. During their flights the bees take up dust and chemicals from the surrounding adsorbing at their hairs. If the bees enter the hive they are controlled by sensors for the specific chemicals. It is planned to built a very small and light radio tags. The tags will be attached to the backs of the bees. So flight direction, flight time and landings could be observed. The researchers believe that this combination will help to pinpoint mines and to reduce areas which are not contaminated by mines.

Another project which focuses on bees is conducted by Sandia National Laboratory [17]. The goal of this project is to train the bees to recognize the smell of TNT and to associate the smell with food. It is assumed that both projects where bees are involved are years away from practical applications.

Antibodies

Antibodies react in a very specific way with single substances, e.g. TNT. The antibodies are mounted on an sensitive piezo-electric quartz microbalance (QMC) system [18]. In the presence of TNT the antibodies attaches itself to this molecules and leave the surface of the QMC which will be accurately measured by a frequency shift of the crystal. In addition the mass of the antibody is 600 times the mass of a TNT- molecule; this leads to a multiplier effect resulting in a much higher sensitivity compared with methods which rely on adsorption of the molecules only. Before the measurement is conducted the TNT is sampled into a solution therefore e.g. 100 liters air are concentrated into 10 ml of a solution. This solution will be examined afterwards. The Advanced Aviation Technology company claimed that their collecting system is able to collect 600 l of air per minute. The company plans to improve the necessary sensor response time to approx. 3 minutes and less. They published that the system is able to measure concentrations down to 10^{-12}, when combined with the concentration process and the multiplier effect this leads to an overall sensitivity of a million times lower at 10^{-19}. This indeed would be a very sensitive system. They planned to develop systems for RDX, PETN and drugs also.

Conducting polymers

At the California Institute of Technology polymer films embedded with conductive materials have been proven as materials for an "electronic nose" [19]. The tested sensors consisting of eight nominally identical carbon black polymer composites. The sorption of a vapor into the polymer films causes physical swelling of the film which leads to a unspecific change in the electrical resistance of the films. As part of a DARPA sponsored program these method was tested to detect and identify chemical signatures of landmines. As a result of another DARPA program laboratory experiments were conducted with DNT instead of TNT. DNT is an impurity that results during the fabrication of military grade TNT. In comparison to TNT the vapor pressure of DNT is much higher (s. Tab.1). In addition in all field experiments concerning the vapor composition above a buried mine DNT was found instead of TNT. The authors set out that in all cases if the sensor is exposed by DNT a signal of the sensors was received and that the actual measured DNT-concentrations are less than 1 ppb. First commercial products based on this method for different chemicals are obviously on the market [20].

Fluorescent Polymers

Within the framework of the DARPA dogs nose program novel fluorescent polymers were developed which are sensitive for airborne TNT [21]. Fluorescence methods are very common for the sensitive estimation of different chemicals. In normal fluorescent measurements the interesting molecule specifically binds to a fluorescent molecule. As a result of the binding the fluorescence is quenched. Therefore the quenching could be used as a measure of the concentrations of the special chemicals. Because of the low TNT concentration in the vapor the quenching is rather low with normal fluorescent molecules. Nomadics, Massachusetts Institute of Technology and the Oklahoma State University developed fluorescent molecules which showed a much higher quenching factor as normal single molecules. Binding of a single TNT molecule can quench the fluorescence of hundreds of polymer repeat units which results in an amplification of the quenching response. The authors stated that a man portable prototype, called FIDO [22], has demonstrated a performance comparable to that of canines during field tests. Two points in their publication are remarkable: first the used polymers showed a higher response for TNT as for 1.3 DNB, 2.6 DNT and 2.4 DNT and second the authors claimed that they measured TNT-concentrations in the headspace over a sample of 56 fg/ml air (6 ppt, n/n).

Surface Acoustic Waves

On top of a piezoelectric substrate an input transducer is placed which launches an acoustic wave that travels over the surface and is detected by an output transducer. The device runs with frequencies up to 100 MHz. The velocity and attenuation of the signal are sensitive to the viscoelasticity as well as the mass of a thin chemical film placed between both transducers. This allows the detection of chemicals specifically bonded at the film. At Sandia National Laboratory a lot of work concerned SAW and substance specific films is conducted. The film used for these SAW's consists of self assembled molecular monolayers that allow the covalent attachment of e.g. TNT- or DNT-specific host reagents. It is stated that using these method low ppb sensitivity is observed for nitroaromates. Moreover it is planned to enhance the sensitivity by 3-4 orders of magnitude [23].

MO_x-Sensors

Today metal oxide sensors are used for the detection and estimation of broad range of different airborne chemicals. They consist of tin dioxide (SnO_2) on a sintered alumina ceramics. To enhance the sensitivity and specificity the SnO_2 is sometimes doped by other metals. In clean air the electrical conductivity is low, while in contact with e.g. reducing gases such as carbon monoxide increases the conductivity. Sensitivity of the sensor may be altered by changing the temperature of the sensing element also. MO_x sensors are sensitive for humidity. As humidity increases the sensor output increases as well.

At the so called Kamina [24] eight metal oxide sensors are combined for pattern recognition. The sensors work e.g. at different temperatures. The sensitivity of the MO_x-sensors is in the ppm-range. It is believed that through using nano-structured materials the sensitivity can be enhanced in a wide range.

Electrochemical Detection

Today electrochemical sensor systems are very common in a wide field for detection of different chemicals. Beside a good selectivity the major advantage is the very high sensitivity of amperometric systems. In addition they are small, easy to handle and cheap compared to some other systems. At amperometric sensors the potential of the working electrode is fixed on one value and the resulting current serves directly as the sensor signal and is normally direct proportional to the concentration of the interesting chemicals. Fig. 1 shows a principal scheme of an electrochemical sensor for vapor detection. Inside the cell body three electrodes are placed. The working electrode is located near or on top of a membrane which serves as interface between the liquid phase inside the body and the vapor phase outside. Chemicals crossing the membrane are electrochemically oxidized or reduced at the working electrode.

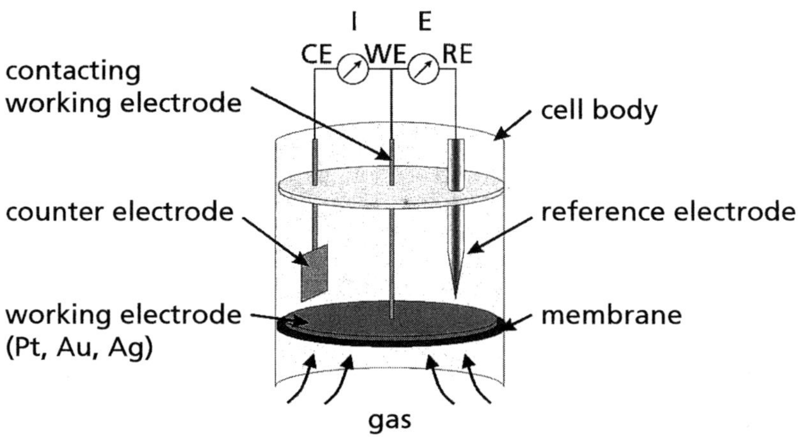

Fig. 1 Principal scheme of electrochemical sensors

The setup of the electrochemical system we develop at the Fraunhofer-Institut für Chemische Technologie varies from this principal setup [26]. The necessary electrolyte is placed directly on top of the tip of the probe, no membrane is used. The reason for that is that up to now we did not find a membrane which is permeable for TNT. TNT adsorbs strongly on most polymer materials and will not cross the membrane. Moreover if the membrane is contaminated with TNT the desorbing TNT disturbs the sensor additionally.

The electrochemical method we use to detect TNT is called cyclic voltammetry, a common electrochemical method which is seldom used for sensors [27]. Fig. 2 shows the potential which is applied to the working electrode. The potential varies in a triangular manner with the time. The potential limits are given by the values of electrolyte decomposition. During one scan the resulting current is recorded as a function of the potential. The shape of the diagram is typical and stable for the combination of the electrode material, here gold, and the kind of electrolyte. If an electroactive substance e.g. TNT is oxidized or reduced the shape of the diagram changes. As can be seen in Fig. 3 a current peak is observed at approx. –0.5 V vs. Au which is typical for nitroaromates. At this potential nitroaromates are electrochemically reduced. We use this signal for a qualitative and a quantitative estimation of TNT from the vapor phase. Up to now we were able to detect TNT concentrations of approx. 500 ppt in the vapor phase. This value is not the lower detection limit but the smallest concentration our vaporizer can deliver so far.

The sensitivity of electrochemical sensors depends additionally on the electrocatalyst used. Experiments with different electrode materials were conducted to estimate the different sensitivities for the TNT detection. As can be seen in Fig. 4 Pt shows only small sensitivity for the TNT detection. The best behavior showed Ag and boron doped diamond (BDD) electrodes. Because of their cross sensitivities against sulfur containing molecules Ag electrodes are not the best choice for the application of mine detection. We plan to use BDD-microelectrodes in the future.

Using Au electrodes measurements were conducted to investigate whether the vapor pressures will influence the sensitivity of the Au electrode alone. These experiments were conducted at the WIWEB. Our normal electrode setup was placed above samples of the single substances, e.g. 3.4 DNT, Nitromethane, 2.6 DAT, C4 etc. In Fig. 5 the peak currents are shown and the vapor pressures of the several substances are given. As can be seen from Fig. 5 the peak currents are independent of the vapor pressures of the

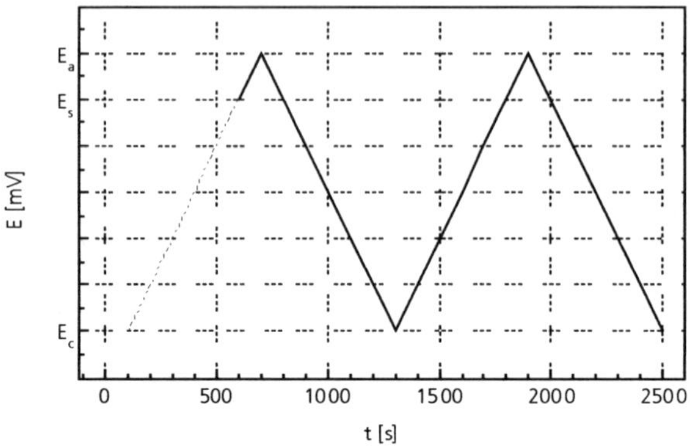

Fig. 2 Potential vs. time course as applied at the working electrode during cyclic voltammetry

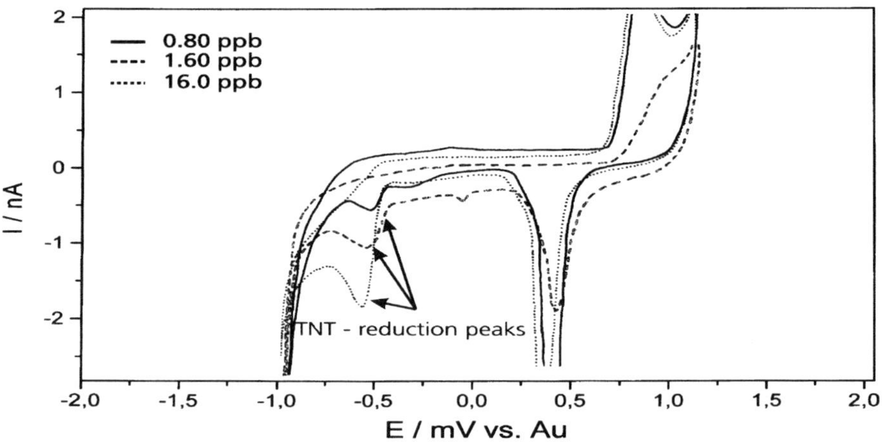

Fig. 3 Cyclic voltammogramms of different vapor TNT concentrations at an Au-microelectrode, v = 100 mV/s

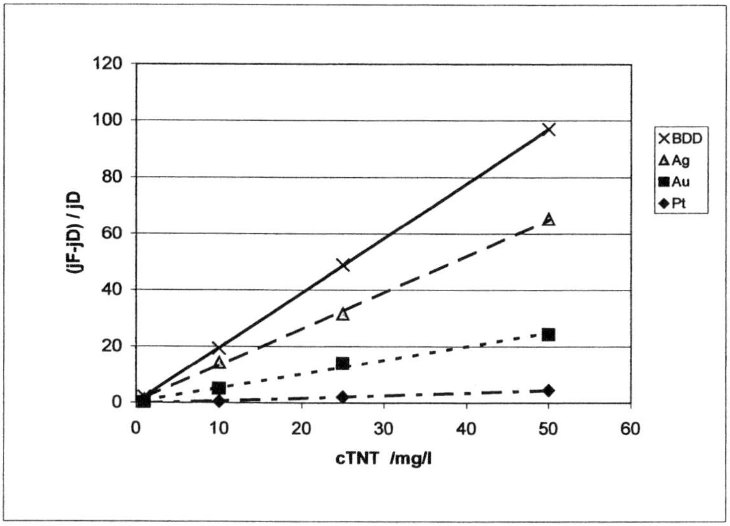

Fig. 4 Electrochemical signal as a function of the TNT-concentration at different working electrode materials measured in liquid electrolyte, v = 100 mV/s

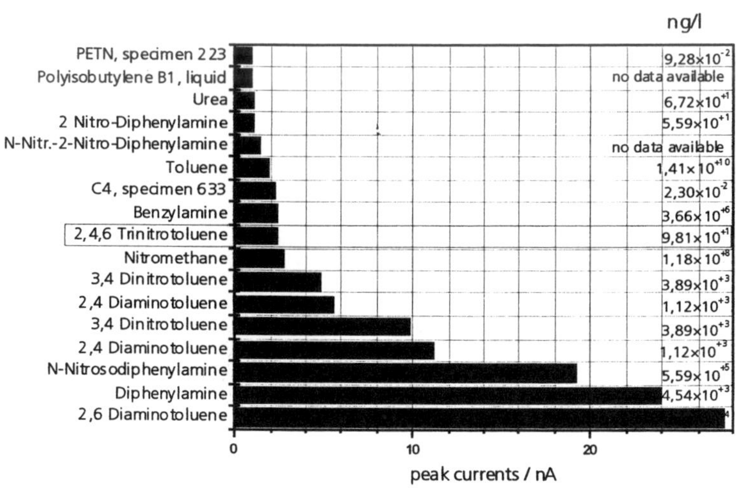

Fig. 5 Peak currents of various substances measured by cyclic voltammetry from the vapor phase compared with the vapor pressures of these substances

substances. Toluene shows the highest vapor pressure (approx. 10^{10} ng/L) of all substances tested but not the highest peak current. The signal height of C4 (10^{-2} ng/L) is nearly the same than the signal of 2.4.6-TNT (10^{1} ng/L). Therefore it seems possible that the sensitivity of the electrochemical TNT-detection might be enhanced by choosing other electrode materials as was presented in Fig. 3 also. Up to now only a few materials were tested for TNT therefore also alloys and other metals should be tested for TNT and in addition for other explosives and drugs.

Conclusions

In comparison to landmine detection by metal detectors and by prodding the use of sniffer dogs is faster. Therefore smelling landmines might enhance the clearing rate. Although dogs are very fast there are some draw backs, e.g. they get tired after some time, they are prone to disease, the combination of handler and the dogs is a complex mixture and training and the support are expensive. To overcome these facts technical solutions are searched far.

Several different chemical, biochemical and electrochemical methods are under study for the purpose of mine detection. There are different methods which are highly sensitive to detect very small TNT-concentrations from the vapor phase. Most of research and development is concentrated on the detection of TNT but analyses of the vapor phase of buried mines and military grade TNT show only very small TNT-amounts, if any, in the vapor phase compared with higher concentrations of DNT and DNB. Therefore it seems reasonable to take these substances into consideration. Both chemicals are contamination of TNT and could be used to locate mines also. Moreover presumably other substances which evaporate from the mines could be used for the chemical mine detection e.g. waxes, softener, etc.. The most sensitive systems could be combined in the way of pattern recognition then. The combination of different methods which focuses on different aspects of a mine seem to be the better way to develop chemical sensor systems.

Mine affected areas vary by climate, soil, vegetation etc. Laboratory experiments could cover a small part of these very variable conditions, therefore information of field experiences, especially from dog handlers are very valuable for the further development of sensors and methods for the chemical mine detection.

Acknowledgements

The financial support of the Bundesamt für Wehrtechnik und Beschaffung (BWB) is gratefully acknowledged. The authors wish to thank the Wehrwissenschaftliches Institut für Werk-, Explosiv- und Betriebsstoffe for their experimental support to measure definite TNT-concentrations and other substances.

Literature

[1] http://www.un.org/cyberschoolbus/banmines/units/unit2b.asp

[2] http://globaled.ausaid.gov.au/secondary/casestud/sri-lanka/1/sri-lanka.html

http://www.hundskolan.se/minedog.htm

http://www.gichd.ch/docs/studies/dogs_4.htm

[3] http://www.mech.uwa.edu.au/jpt/demining/k9/dog-measure.pdf

[4] http://www.angola.npaid.org/mines_database.htm

[5] Rudolf Meyer, Explosivstoffe, VCH Weinheim, 1985

[6] http://www.vetmed.auburn.edu/ibds/frame.htm

[7] http://www.vetmed.auburn.edu/ibds/frame.htm

[8] D.C. Leggett, T. F. Jenkins, A. W. Hogan, T. Ranney and P. H. Miyares, "External contamination of Landmines by organic nitro-compounds", US Army Corps of Engineers, Technical Report ERDC/CRREL-TR-00-2, March 2000

[9] T. F. Jenkins, D. C. Leggett and T. A. Ranney, "Vapor signature from Military Explosives", US Army Corps of Engineers, Special Report 99-21, December 1999

[10] V. George, T.F. Jenkins, J.M. Phelan, D.C. Leggett, J. Oxley, S.W. Webb, P.H. Miyares, J.H. Cragin, J. Smith, T.E. Berry,

"Progress on determining the vapor signature of a buried landmine",

in Detection and Remidiation Technologies for mines and minelike Targets V, Proceedings of SPIE Vol. 4038 (2000)

[11] A. H. Kjellström and L. M. Sarholm,

"Analysis of TNT and related compounds in vapor and solid phase in different types of soil",

in Detection and Remidiation Technologies for mines and minelike Targets V, Proceedings of SPIE Vol. 4038 (2000)

[12] S. W. :Webb and J.M. Phelan, "Effect of diurnal and seasonal weather variations on the chemical signatures from buried landmines/UXO" , in Detection and Remidiation Technologies for mines and minelike Targets V, Proceedings of SPIE Vol. 4038 (2000)

[13] F.P. Karg and G. Koss, "Untersuchung zur Umweltchemie, Kontaminationserkundung und –bewertung", Z. Umweltchem. 5 (1993) 182

[14] http://users.skynet.be/apopo/rat.htm

[15] http://www.mgm.org/e/research/rats_1.htm

[16] http://www.pnl.gov/news/1999/99-13.htm

[17] http://www.sandia.gov/media/minebees.htm

[18] http://www.bioapp.se/eng-tekprod_intro.html

http://www.aatl.net/publications/Bio-sensor.htm

[19] http://www.siam.org/meetings/sdm01/pdf/sdm01_11.pdf

[20] http://www.chipcenter.com/eexpert/lgoldberg2/tbreunig001.html

[21] http://www.nomadics.com/Landmine_Detector/Brochures_white_papers/uxo2001.pdf

[22] http://www.nomadics.com/Landmine_Detector/Fido/fido_develop_status.htm

[23] C. K. Ho, M.T. Itamura, M. Kelley and R.C. Hughes, "Review of chemical sensors for in-situ monitoring of volatile contaminants", Sandia Report SAND2001-0643, March 2001

[24] http://irchsurf5.fzk.de/mox-sensors/Default_Eng.htm

[26] M. Krausa and K. Schorb, "Trace detection of 2,4,6-trinitrotoluene in the gaseous phase by cyclic voltammetry", J. Electroanal. Chem., 461 (1999) 10

[27] C. H. Hamann and W. Vielstich, Electrochemistry, Wiley-VCH Weinheim, 1998

CONCEPT OF A COMBINED MOBILE DEVICE FOR EXPLOSIVES AND LANDMINES IDENTIFICATION BASED ON TIMED NEUTRON SOURCE AND ELECTROMAGNETIC UHF WAVES

A. KUZNETSOV
V.G. Khlopin Radium Institute
2-j Murinski pr., 28, 194021 St Petersburg, Russia

1. Introduction and short overview of explosives substance' detection methods

Detection of modern explosives substances (ES) that contain little or no metal is a demanding task. Different methods have been proposed, that can be split into two broad categories:

- Localization of "anomalies" which could be hidden ES;
- Identification of ES.

Localization methods are used to quickly find an "anomaly" in the inspected region or object. Identification methods are used to find ES among many anomalies (empty cavities, stones, roots, metallic or other dense materials, light polymer compounds or hidden explosives).

The most common localization methods are:

1. detecting presence of metal coating or detonator (metal detectors);
2. using an electromagnetic radar, acoustic or infra-red detectors;
3. methods based on neutron, gamma, or X-ray transmission and back-scattering from studied object.

The most common identification methods are based on:

1. the analysis of ES vapors (trained dogs, chromatographic analysis, laser-based sensors of characteristic absorption bands in infra-red range, "electronic nose" based on measurements of conductivity of thin sensors of tin dioxide, X-ray fluorescence analysis of nitrogen-containing molecules, etc.);
2. detection of high concentration of nitrogen which is typical for most ES (irradiation with thermal neutrons with detection of the characteristic gamma-line 10.8MeV – TNA; nuclear quadruple resonance of nitrogen atoms observed at radio frequencies for some ES);
3. determination of the elementary composition, which is typical for ES - relation between concentrations of carbon, oxygen, nitrogen, hydrogen – by irradiation with fast neutrons and subsequent identification of characteristic gamma-lines of the main elements (Fast Neutron Analysis - FNA).

Taking into account advantages and disadvantages of the above explosives' detection methods and having in mind the existing need in an inexpensive mobile device for detection (identification) of ES, we have concentrated our research at

H. Schubert and A. Kuznetsov (eds.), Detection of Explosives and Landmines, 21–32.

Khlopin Radium Institute (ISTC Project #1050) around the development of two methods of explosives' detection:

- application of electromagnetic UHF waves for localization and preliminary identification of hidden objects by using their dielectric properties
- irradiation of the localized object with neutrons and subsequent determination of the elemental composition of the object

2. Localization and preliminary identification of explosives using electromagnetic UHF waves

To investigate dielectric properties of substances and to carry out reliable modeling of processes of interaction of microwave radiation with real objects in a wide range of frequencies the automated experimental installation was created. The ranges of frequencies of the microwave radiation (5-8 GHz, 8-12 GHz, 17-25 GHz, depending on the generator option) were chosen to satisfy two conditions: high spatial resolution (1,2 - 6cm) and acceptable penetrating ability of radiation in damp and wet medium. Unlike pulsed radars, in our method the inspected area is continuously irradiated by a probing wave with changing frequency. The power of the radiation is ~1mW, and the dynamical range of the existing receiver is 40dB. During the panoramic measurements the frequency is determined with accuracy about 0.1-0.3 %.

To determine dielectric properties of explosives substances and their distinction from other hidden objects the following explosives were investigated:

- TNT; trinitrotoluol), $C_7H_5N_3O_6$;
- (RDX; cyclonite), $C_3H_6N_6O_6$;
- AIX-1 (94,5% - RDX, 4,5% - inert compound);
- Trinitrophenylmethylnitramine, tetryl: $C_7H_5N_5O_8$.

Measuring UHF waves reflected from an object located under the layer of ground at various depths we were able to localize and carry out preliminary identification of hidden objects by their dielectric properties. Geometry of experiment is shown on Figure 1.

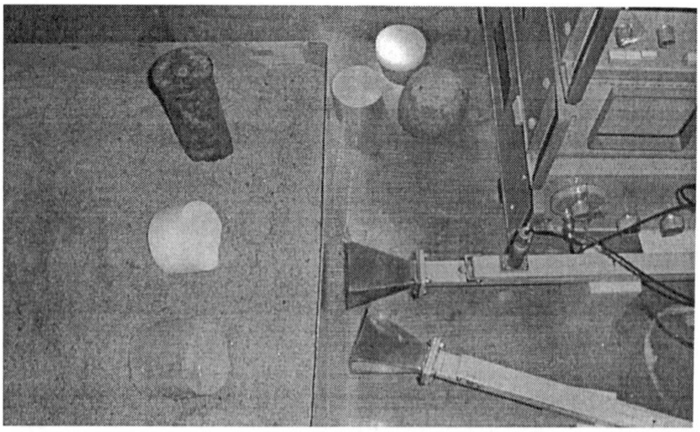

Figure 1. Photo of the laboratory test area for UHF technique. Samples of different materials – metal shell, stone, TNT imitator, wood – can be seen at the left-hand side of the photo.

Samples were placed in sand with humidity 0.1 and 1 % by weight. Investigated objects were:

- imitator of TNT (trinitrotoluol) – cylinder 70mm in dia. and 40mm height
- wet root - cylinder 60mm in diameter and 60mm height
- metallic cylinder 50mm in diameter and 100mm height
- metallic disc 70mm in diameter and 10mm height
- stone with dimensions 110x70x75mm

While scanning the surface of sand at 2cm/sec, frequency dependence of attenuation of reflected signal with respect to the primary one in the frequency range 5-25 GHz was investigated. Dependence of power attenuation of reflected signal in the frequency range 8-12 GHz for three hidden objects located under 7cm layer of sand is shown on Figure 2. The same dependence in the range 17-25 GHz for three different hidden objects is displayed on Figure 3. Figure 4 shows Fourier transformation of the dependence displayed on Figure 3. Analysis of spectra on Figure 2 allows one to localize and determine dimensions of the object. Analysis of Fourier transformation allows one to determine the tenancy of the object to various groups: metal, root, other dielectric (stone or imitator of TNT). Measurements with dry sand (with humidity 0.1% by weight) showed that the maximal depth of the object is 20cm. This limit is mostly due to low sensitivity of the receiver and low power of the probing wave. Measurements with wet sand (humidity 1% by weight, which correspond to about 5mm of rain) allow to carry out preliminary identification of hidden objects at depths up to 7cm. This value can be increased by increasing the sensitivity of the receiver (40 dB is current limit of the laboratory installation) and increasing the power of the probing signal(1mW for current laboratory installation).

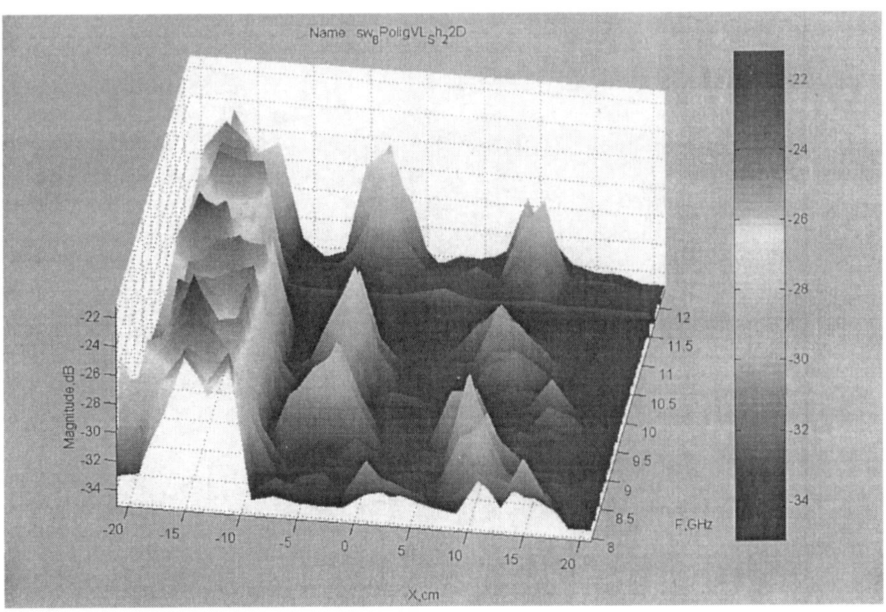

Figure 2. Dependence of power attenuation of reflected signal in frequency range 8-12 GHz for three hidden objects located under 7cm layer of sand: metallic cylinder (location coordinate -13cm), TNT imitator (location coordinate -3cm), wet root (location coordinate +9cm).

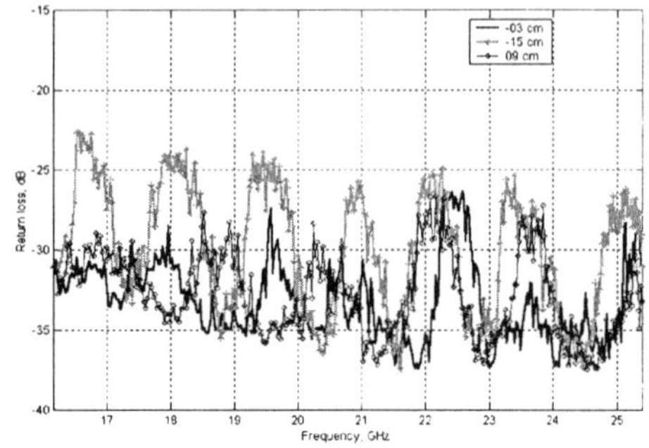

Figure 3. Dependence of power attenuation of reflected signal in frequency range 17-25 GHz for three hidden objects located under 7cm layer of sand (metallic cylinder , TNT imitator, wet root).

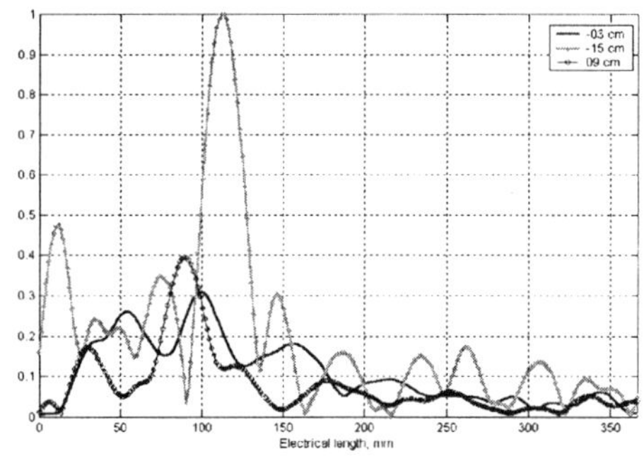

Figure 4. Fourier transformation of the spectra shown at Figure 3.

On the basis of laboratory tests that have been done so far, the following conclusions can be made:

- An original technique for detection of objects, which are located behind the air-sand border, by detection of reflected UFH-range electromagnetic waves was developed. The technique allows to carry out not only localization of an object and determination of its dimensions with accuracy 0.5cm, but also to establish tenancy of the object to a certain class: explosive substances with small metal content, metal objects, roots of trees, stones, etc. The parameters of laboratory installation allow one to use it at thickness of a ground up to 20cm for dry sand, or up to 7cm for wet sand. The speed of scanning is determined by the period of scanning the whole frequency range, which is currently makes 2cm/s for the laboratory installation.
- This technique does not allow one to distinguish between metal objects containing explosive substances and metal objects without explosives, but can signal the presence of metal in the inspected area.

3. Technique for identification of explosives based on the timed neutron source

For identification of the located anomalies we use a technique based on irradiation with neutrons from timed isotopic source with the subsequent determination relative concentrations of basic elements (carbon, oxygen, nitrogen, hydrogen) in the inspected object.

The main idea for modernization of the technique based on an irradiation with neutrons from radioactive sources (for example, ^{252} f) is to simultaneously measure secondary gamma-radiation from the object and accompanying (charged) particles. The latter allows to use «marked» neutrons (i.e. to determine, from which reaction the given neutron originates), and to carry out detection of secondary gamma-radiation in narrow (nanosecond) time intervals, and thus to considerably improve background conditions.

The accompanying particles (fission fragments of ^{252}Cf) are detected by a specially developed ionization chamber with ^{252}Cf source inside. A small chamber with a 2 μg ^{252}Cf source (~5×10^6 neutrons per second) is shown on Figure 5. The source was made by the micro-drop method with subsequent high-temperature annealing. Before the process, the ^{252}Cf material was purified by the distributive chromatography method. Then the material was micro-dropped on the polished platinum disk with diameter 24mm and thickness 0.3mm, which is located at 3cm from the collector plate. The diameter of the ^{252}Cf spot was 12mm. For prevention of self-dispersion of substance a thin (120 μg/cm2) layer of gold was evaporated on the surface.

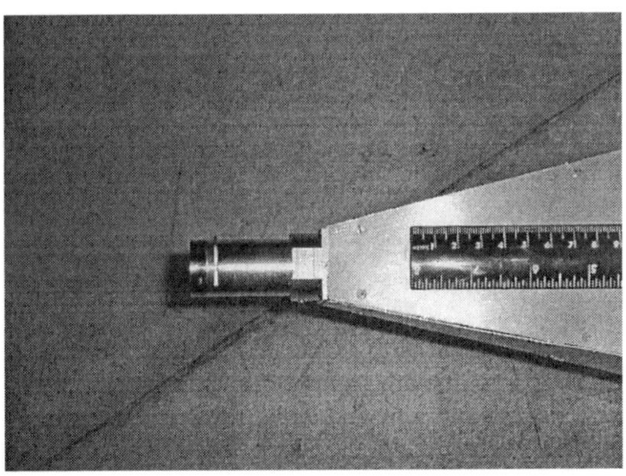

Figure 5. Ionization chamber with 2μg ^{252}Cf source and preamplifier coupled to a pyramidal shielding, consisting from layers of iron, polythene and lead.

Gamma-rays are detected by a NaI(Tl) crystal with dimensions: diameter 150mm × 100mm coupled to a photo multiplier. The detector is placed on a mobile platform (see Figure 6). The mobile installation is serviced by an electronic block, which was specially developed for use in various temperature and weather conditions. It includes programmable microprocessor, PC-connection, spectrometer with possibility to use nanosecond windows of gamma detection, and a battery power supply. Hardware and software allow to contineously monitor count rates of the ionization chamber, gamma-detector, as well as the rate of coincidences. Temperature effects on the spectrometer are corrected on-line by altering the amplification in spectroscopic channels.

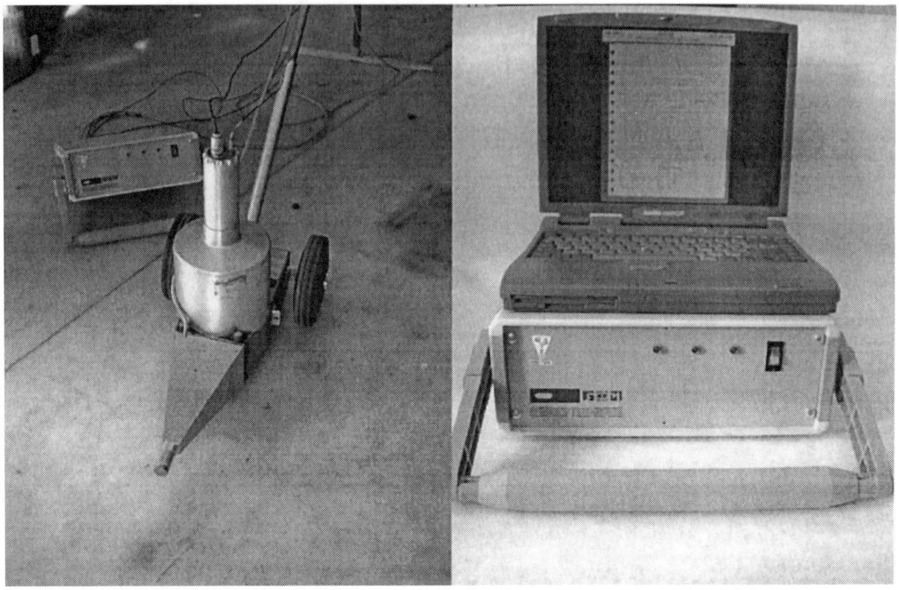

Figure 6. A prototype of the mobile device for detection of explosive substances.

Analysis the gamma-spectra shows, that introduction of time coordinate (selection of a narrow interval of time of the order 10 ns for detection of gamma-rays) has allowed to reduce background connected with direct fast neutrons and gamma-rays from the source by more than an order of magnitude, and also to reduce background from cosmic radiation, thermal neutrons and other uncorrelated background by two orders of magnitude. This in turn allows one to identify characteristic lines of main elements: oxygen ^{16}O (6.13MeV), carbon ^{12}C (4.43MeV), and nitrogen ^{14}N (2.31 and 5.11MeV).

The process of identification of explosive substances was simulated in a computer code based on Monte-Carlo technique. Calculations include modeling the source of neutrons, real geometry of the gamma detector, response functions of the detector. Calculations reproduce time dependence of the spectrum of gamma-rays and neutrons, and also distribution of pulses in gamma detector.

Measurements of spectra secondary gamma-rays were carried out using objects placed either in open air, or under a layer of a soil of various thickness, composition and humidity. The following types of soil were used: fertilized soil with humidity up to 20 % by weight, and sand with humidity up to 0.4 % by weight. Measurements for the following samples were done:

TNT phantom (with relative concentration of carbon, oxygen, nitrogen, and hydrogen similar to those of the real TNT) of cylindrical shape weighting 500g and 700g.

- Damp root of a tree - cylinder with weight about 700 g.
- The steel cylinder with weight 700 g.
- Stone weighting about 900 g.

The view of the experimental area is shown on Figure 7.

Figure 7. Geometry used in experiments with the prototype of mobile device for explosives detection with 2 μg ^{252}Cf timed neutron source.

Samples were located at distances from 2cm to 5cm below the surface of the soil. Time of measurements varied from 5 to 20 minutes depending on the depth and on the required probability of identification.

Experimental spectra of secondary gamma-rays obtained in these experiments are shown at Figure 8.

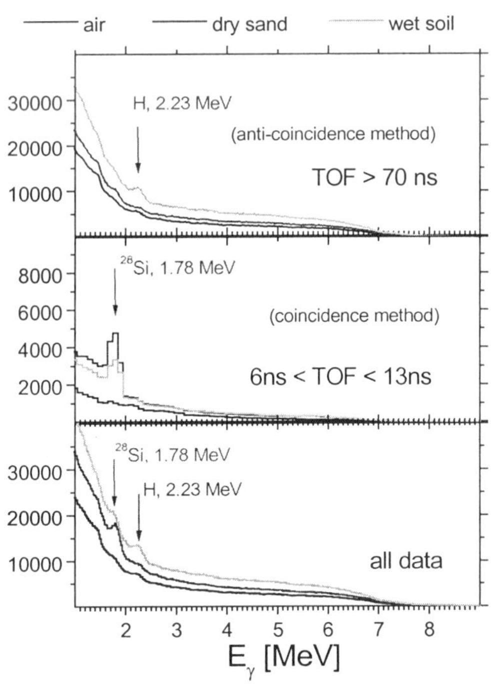

Figure 8. Experimental spectra of secondary gamma-rays in air, damp soil and dry sand for various time windows. Collection time – 10 minutes.

It is clearly visible from the Figure that selection of narrow time interval of the order of 8ns for gamma-ray detection leads to a decrease of the background by more than an order of magnitude. Due to efficient suppression of the background associated with thermal neutrons, backgrounds from different soils with different water content become similar to each other in the interesting area of energies (see the middle plot on Figure 8).

Response functions of the existing gamma-detector to gamma-rays from reactions of neutrons with main chemical elements and their compositions were simulated in the computer code based on Monte-Carlo technique. Geometry of the neutron source, gamma-detector and construction materials were taken into account. Some of the response functions were measured experimentally for comparison with experiment. In all cases, experimental and simulated response functions coincide within uncertainty limits.

We are currently using a decision taking procedure that relies on principal component analysis with Mahalonobis distances (PCA/MD). This method allows one to obtain probability that the unknown spectrum belongs to the given group of spectra from the training set. The main advantage of the method as compared to the least-squares fitting procedure is that it accounts for common variations of spectra at the given energy as well as correlations between different energy points of the spectrum (i.e. co-variation). In other words, practically all the information contained in the spectrum is taken into account.

The PCA/MD method of identification requires several measurements of the background conditions at different locations around the inspected object. These measurements, together with results of the Monte-Carlo calculations, allow one to build a library of training spectra, which are then used by the system for recognizing the unknown substances by their gamma-spectrum.

The analysis of experimental spectra shows, that this technique allows us to identify explosive substances weighting around 500 g within 5-15 minutes (depending on the depth and type of soil) with probability 99 % at 1 % of false alarms. The results of identification are not sensitive to presence/absence of metal, and can reliably distinguish explosives from nitrogen-containing organic substances (tree root, melamine, etc).

4. Technique of identification of explosive substances using neutron generator with nanosecond timing based on detection of the accompanying α-particles

The most important disadvantages of using the timed isotopic neutron source are: a) the need to use heavy shielding against radiation during storage and transportation of the equipment; b) relatively low flux of fast neutrons, which is limited by the radiation safety rules. To achieve better characteristics of the current prototype one has to use different types of timed neutron sources, for example neutron generator with nanosecond timing. Unlike isotopic source, a portable neutron generator can be switched off for storage and transportation. Detection of α-particles which accompany neutron emission in T+D reaction allows one to collimate neutrons flux on the suspected location of the hidden object and to identify it through the analysis of spectrum of secondary gamma-rays.

We have carried out model calculations of the parameters of the device that uses a portable neutron generator with total neutron flux 10^8 neutrons/second with and without the system for detection of the accompanying particles.

The results of these calculations are presented on Figures 9 and 10.

Explosives detection using portable neutron generator.

Set-up parameters

Neutron flux	- 10^8 neutron/sec.
Eplosive	- TNT 85 gramm under 4 cm of soil surface, 30 cm away from crystall
Gamma detector	- 7.5 in dia. x 7.5 cm BGO crystall
Passive shield	- Tungsten, 40 cm cone.

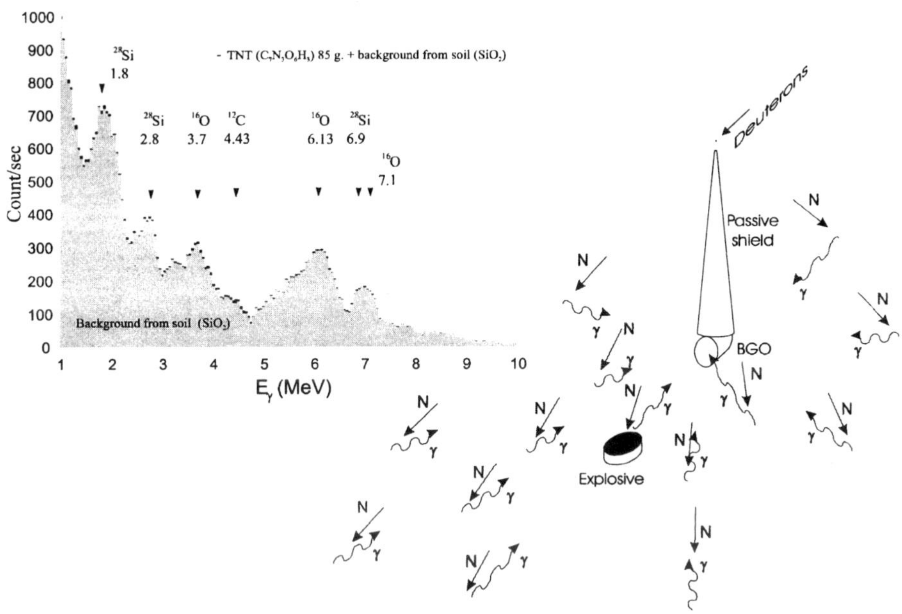

Result of Monte-Carlo calculations

BGO count rate - 118 207 count/sec

For energy window 3.8 - 4.8 MeV (^{12}C line 4.43 MeV)

Count rate for explosive - 3.5 count/sec

Time of identification of ^{12}C line 4.43 MeV

(2σ over background): 385 sec

Figure 9. Calculated detection parameters for 85 g TNT located under 4cm of sand by a device with D-T neutron generator with 10^8 n/s neutron flux without detection of the accompanying particles.

Explosives detection using portable neutron generator with associated α- particles detection.

Set-up parameters

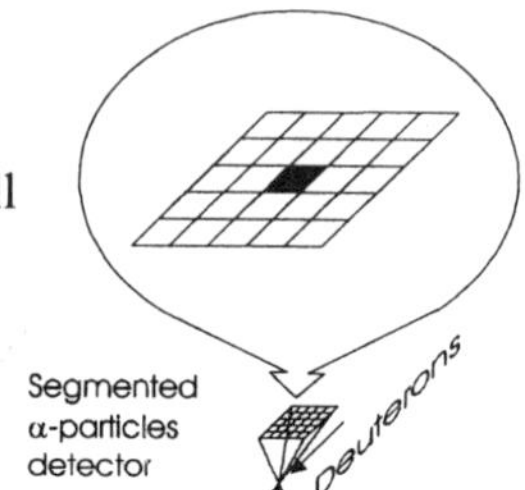

Neutron flux - 10^8 neutron/sec.
Eplosive - TNT 85 g. under 4 cm of soil surface, 30 cm away from crystall
Gamma detector - 7.5 in dia. x 7.5 cm BGO crystall
Passive shield - Tungsten, 40 cm cone.
α-particles detector - 2.5 x 2.5 cm^2, with 25 segments.
Investigated area - 50 x 50 cm^2.

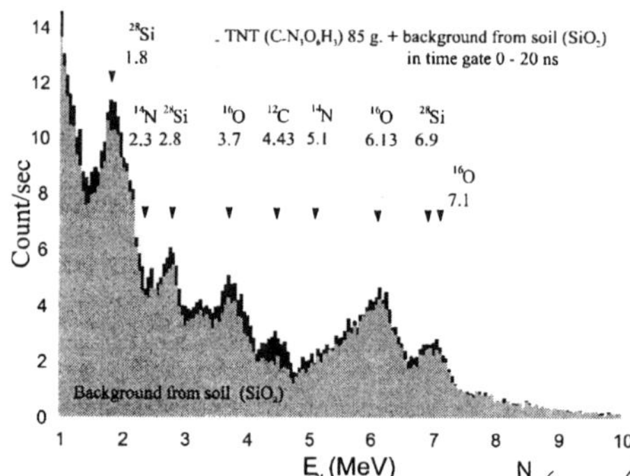

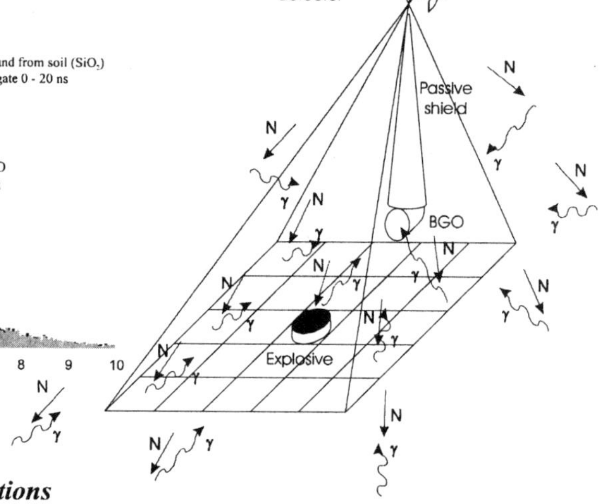

Result of Monte-Carlo calculations

BGO count rate - 118 207 count/sec
α-particles count rate per segment - 320 000 count/sec
Accidental coincidence - 99 756 count/sec in time gate 0-10 μs
900 count/sec in time gate 0-20 ns
For energy window 3.8 - 4.8 MeV (^{12}C line 4.43 MeV)
Count rate for explosive - 3.5 count/sec
Accidental coincidence - 1 133 count/sec in time gate 0-10 μs
9 count/sec in time gate 0-20 ns
Time of identification of ^{12}C line 4.43 MeV (2σ over background):
385 sec in time gate 0-10 μs
4 sec in time gate 0-20 ns

Figure 10. Calculated detection parameters for 85 g TNT located under 4cm of sand by a device with D-T neutron generator with 10^8 n/s neutron flux and system for detection of the accompanying particles with geometrical efficiency ~8%.

The sectioned detector of the accompanying α-particles should possess the following characteristics:

- Have intrinsic detection efficiency ~100% for α-particles;

- Have sub-nanosecond time resolution;
- Must be capable of working at high count rates (up to 10^7 particles/second).

To meet the above requirements, we constructed a semiconductor detector and tested it at the NG-400 neutron generator of the Applied Physics Laboratory at Radium Institute. The detector was exposed to an integrated doze of neutrons and α-particles 10^{13}, which corresponds to about 500 hours of operation of the portable neutron generator. The detector remained operational throughout the irradiation.

The prospective characteristics of the prototype of the device for explosives detection equipped by neutron generator with nanosecond timing:

- Identification of the hidden object weighting 100 g within 10 seconds with probability >99% and probability of false alarms <1%.
- Simultaneously inspected area 50cm × 50cm, position resolution within this area – 10cm × 10cm.
- Time of preparation of the device - around 5 minutes. Automated controls and the decision-taking procedure ("single button" approach) will not demand special knowledge from the personnel.
- Mobility: weight of the device - not more than 40 kg, dimensions: 50×40×30 ^{3}.
- The installation will comply to the existing radiating safety rules and will not cause damage to health of people, goods and vehicles.
- Low maintenance cost - low power consumption ~20 Watt, estimated life time without replacement of the neutron source ~ 2000 hours.

5. Further development of the prototype of the mobile device for detection and identification of explosive substances

Our experience of laboratory measurements in soil and sand of various humidity shows that the two techniques are complimentary to each other and allow reliable identification of ES. For example, while the UHF technique may have difficulties in distinguishing between TNT and stone, which have similar dielectric properties, such distinguishing can be easily achieved with nuclear technique. The nuclear technique, in its turn, may not be able to reliably distinguish TNT from a wet wood (e.g., tree root) in some environments, while for a UHF method this is a trivial task, since these materials differ strongly in water content.

Further development of the existing prototype of the mobile devise for explosives detection will be done along the following lines:

- Creation of the prototype of the portable device for localization and preliminary identification of explosive substances by electromagnetic waves of UHF range. Increase of the power of the emitted wave and of the sensitivity of the receiver in order to be able to work in damp soils and at greater depths.
- Laboratory and field tests of the prototype of mobile installation for detection of explosive substances with 2μg ^{252}Cf timed neutron source.
- Development of the decision-taking procedure:
  - Expansion of the library of training spectra at other types of explosives, types of soil, background conditions, etc.
  - Further development of the method on a basis PCA/MD.

- Inclusion of the portable neutron generator with position-sensitive detector of accompanying particles (α-particles) into the existing prototype of the mobile device, which would allow us to reduce time of identification down to several seconds and to obtain 3D image of the inspected area in terms of elemental content for the purposes of localization and identification of the hidden object.

Literature

Kuznetsov A.V., (1999), “Development of a Spontaneous Portable Neutron Source with a Built-in System for Registration of Accompanying Particles for Background Reduction” // In IAEA Research Coordination Meeting “Application of Nuclear Techniques to Anti-personnel Landmines Identification”, 23-26 November 1999, RBI, Zagreb, Croatia.

Kuznetsov A.V., (2000), “A Spontaneous Portable Neutron Source with a Built-in System for Registration of Accompanying Particles for Background Reduction”// In Proc. of The International Conference “Explosives and Drug Detection Techniques”, 18-24 June, 2000, Crete, Greece.

Kuznetsov A.V., (2001), “A Prototype of Mobile Device with Isotopic Timed Neutron Source”// In Proc. of IAEA First Project Co-ordination Meeting on Regional Technical Co-operation Project RER/1/005 “Field Testing and Use of Pulsed Neutron Generator for Demining”, 12-14 February, 2001, Vienna

Kuznetsov A.V., (2001), “Nuclear sensor for explosives detection based on timed radioactive source” // In Proc. of the Second Australian-American Joint Conference on the Technologies of Mine Countermeasures conference, 27-29 March, 2001, Sydney, Australia.

Kuznetsov A.V., (2001), “A prototype of mobile device for explosives and landmines identification based on timed radioactive source”// In Proc. of the 7th International Conference on Applications of Nuclear Techniques "Nuclear and Atomic Industrial & Analytical Applications", 17-23 June, 2001, Crete, Greece.

DETECTION OF NANOGRAM AMOUNTS OF EXPLOSIVES

BORIS VETLICKÝ
(University of Pardubice, Czech Republic)

The detection of small amounts of explosives in the form of small particles or their vapours is permanently an actual theme in the effort of jeopardy prevention of important persons, objects and means of transport by terrorist actions carried out by use of explosives. The recognition of very small amounts of explosives provides also certain possibilities of the detection of the ammunition inclusive of mines, using standard explosives, if however the surface of the ammunition was not especially cleaned after the assembly.

In present time the detection of explosives is being carried out by use of a couple of different ways, each of them has its own selfhood – I do not use the expressions advantages and disadvantages advisedly. These methods are sufficiently known, therefore I shall be very compendious:

1 biological detection – trained dog
2 colour chemical reactions
3 detection of ionised explosives
4 chromatographic methods

ad 1) The detection by use of t rained dogs. An indisputable advantage of this method is virtually the highest sensitivity of dogs to know them or selected explosives and relatively operative possibility of putting on in the case of suspicious object matters. An other question is that the dog is not a chemist, he is able to remember certain hazardous smell, however the smell of the same explosives manufactured by different manufacturers (for instance TNT) may be different. Also the time of continuos (intensive) activity is limited. This is well known from the criminology and blood sports too.

ad 2) The detection of the particles by use of colour chemical reactions. It is necessary, above all, to acquire a sample of particles by a convenient method (by the abrasion of the body, luggage, suit). After acquirement of the sample it is possible to determine on the basis of colour chemical reactions relatively

H. Schubert and A. Kuznetsov (eds.), Detection of Explosives and Landmines, 33–37.

accurately the type of explosive by use of reagent liquids of the detection set. The whole procedure is pretentious in time indeed and requires specific chemical knowledge too.

ad 3) The methods based on the mobility of ionised molecules of explosives in magnetic field. These methods provide the possibility of relatively quick identification of a group of substances with an identical or close molecular mass and ionisation potential. This results in a very quick operative signal of the hazard. Exact identification of the explosive is the further stage of the investigation.

ad 4) Chromatographic methods. These methods enable exact identification of explosive compounds. They are, however, pretentious in time and convenient rather for laboratory analysis. The contemporary trend of mass decrease of this equipment is logic. However, it was yet not possible to reach the low mass of instruments based on the mobility of explosives ions. The time factor is at the operative work not negligible too.

Experimental part

At our University of Pardubice, department of the Theory and Technology of Explosives the following problem was studied: under which conditions would it be possible to detect the least amount of actual explosives by use of the technique disposable in present time, with contingent small structural improvement, leading to the possibility of explosives detection in nanogram amounts or less.

We pursued the compounds having the tension of vapours as low as possible, i.e. most complicated events. For our work we came out of a very objective publication [3], which I reported at the ARW in Prague in the year 1997 and of further literal data [4, 5]:

Our target was the detection of explosives with a minimum vapour tension i.e. of hexogene and octogene.

From basic physical principle follows, that upon the temperature decrease the amount of explosive molecules in the air will decrease, at the temperature increase this amount will be higher too. The functional dependence of the vapour tension upon the temperature confirms a series of publications cited in the part "Literature ".

In our experiments we prepared the specimens of microgram and later of nanogram amounts of shown explosives in gaseous state at various increased temperatures respectively in a small volume of 100 up to 200 ml, to avoid the leakage of molecules into the surroundings. By this way we obtained a specimen sufficiently concentrated for the detection. The experiments were carried out with the use of a series of apparatus, based on the principle of nonlinear mobility of ions in a magnetic field NLDM, from the original M-01, M-2 and more new MO-2 up to MO-2M, where our previous opinion was utilized.

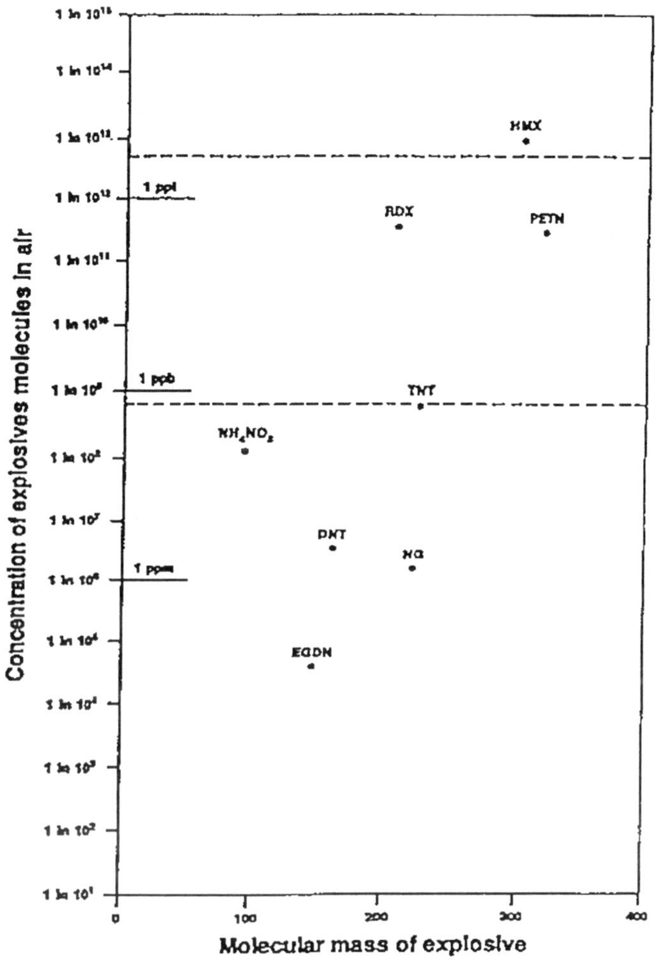

Obtained results confirmed, that a sample of low volatile explosives, acquired by abrasion with following warming enables to advice reliably even insignificant amounts of explosive. This has a special importance for the detection at very low temperatures. – in the case of acquirement of particles by abrasion the result is practically independent of the temperature. The detection apparatus should of course work in the prescribed regime in a room, car etc.

An other variant being verified is based on the accumulation of molecules of explosives in a very distracted milieu, for instance in rooms, with use of special but very simple equipment as a small handy exhauster powered by a battery. Sufficient is the medium (vehicle), where the molecules are picked up and after subsequent warming liberated for the detection. The principle is naturally well known, was however used above all for large detectors of chromatographic type. In our case the detection by a portable handy detector is feasible. The advantage of this system reclines upon the fact, that the taking of samples can be carried out by use of a simple apparatus and by unqualified and only informed workers. Only the evaluation of samples should be carried out by one qualified operator.

Application of this version is effective in the case of solving of a possible diversion in large buildings (public offices, banking houses, transporting centres, hospitals), where the time is a very significant factor. Even if most of these alarms are false but they must not be undervalued. In our view it is possible to contribute significantly by use of relatively operative methods to the detection of the ammunition, containing classic explosives TNT, RDX, HMX, which has not been specially treated on the surface after the assembly and despoiled of insignificant particles of explosives. Such treatment is, however, unreal in the conditions of working with large amounts of ammunition.

At the conclusion of my article I should like to refer two themes which deserve, on our opinion, further elaboration.

a) marking of plastic explosives in accordance with the international convention can be carried out by use of a series of compounds, given in the appendix. We suppose, however, that it is possible too to apply a very simple compound, whose properties are known during long years and which could be determined for certain by use of contemporary techniques in very small amounts. This compound is TNT. We should like to please this forum to consider this possibility and eventually to stimulate the replenishment of TNT as a marking compound.
b) In the course of some experiments, where I was personally present, was found a prompt of the fact, that in the systems of solution type or only in a mixture of explosives with more volatile compounds (solvents, oils etc.), occurs during the evaporation of the medium the entrainment of explosive molecules in amounts sufficient for the detection. It is naturally self-evident, that these signalling information should be scientifically studied and verified. Thy could open some new directions in the science of the detection of explosives. I should appreciate your standpoint in these questions.

Literature

1) NATO ARW Workshop, 11. - 13. November, 1997, Praha.
2) International Symposium, 10. - 11. September, 1996, Pardubice
3) A.Feinberg, Explosive Detection for Aviation Security, Science 225, 1531 – 1537, 1992
4) R.B.Cundall, T.F.Palmer, C.E.,C.Wood Vapour Pressure Measurement on Some Organic High Explosives, 1997,
5) D.C.Dionne, D.P.Rounbehler, E.K.Achter, J.R.Hobbs, D.H.Fine, Vapour Pressure of Explosives, Journal of Energetic Materials, Vol. 4, 447-472, 1986
6) 6th International Symposium on Analysis and Detection of explosives, July 6-10, 1998, Prague, Czech Republic

Humanitarian demining: sensor design and signal processing aspects

M. Acheroy, P. Verlinde
acheroy@elec.rma.ac.be, patrick.verlinde@tele.rma.ac.be

Royal Military Academy (RMA)
Avenue de la Renaissance 30
B-1000 Brussels - Belgium
http://www.sic.rma.ac.be

10th of September 2001

Abstract

This paper presents the aspects of sensor design and signal processing in the context of the HUDEM[1] project. The HUDEM project, dedicated to humanitarian demining, is a nice example to treat, because of the fact that a lot of different technologies are explored. The contribution of this paper is to describe the (results of the) efforts to increase our knowledge on sensors and sensor/ground characteristics, on designing new sensors or fine-tuning old ones, and on processing the data produced by sensors. Also, in this project the detection is also considered as a global process in which the outputs of the sensors, considered as skilled specialists, are integrated in a fusion operation.

1 Introduction

Currently, about 60 million anti-personnel (AP) mines are polluting the environment in about 60 countries. Because mine clearing operations proceed much slower than mine laying ones, the number of polluting mines is still increasing. Humanitarian mine clearance operations must be understood and designed correctly, keeping in mind that their main goal is to provide efficient aid to innocent people, who may be severely injured by this dreadful pollution.

Furthermore, the analysis of actual demining campaigns primarily reveals the far too long time needed to clear polluted terrain, a far too large false alarm rate, the threat of plastic mines (which are difficult to detect by classical means i.e. by metal detectors), and the large variety of mine clearance scenarios, depending on the country, the region, the climate and the place of the pollution (houses in villages, roads, agricultural fields, etc).

The important parameters which characterize the mine detection problem are the mine occurrence probability, the detection probability of a given material and the false alarm probability of a given material [1]:

- The *mine occurrence probability* in a given position of a minefield expresses the local mine density of that minefield as well. Obviously, it is impossible to control this parameter because

[1]The HUDEM project is the Belgian project on Humanitarian demining and is funded by the Belgian Ministry of Defense and by the Belgian State Secretariat for Development Aid. In the HUDEM project, eight Belgian universities are collaborating under the direction of the Royal Military Academy

H. Schubert and A. Kuznetsov (eds.), Detection of Explosives and Landmines, 39–56.

it depends on the reality of the terrain. Nevertheless, this parameter is very important for assessing the probability of an alarm in a given location of the minefield.

- The *detection probability* is the probability of having an alarm in a given position of a minefield for a given detection material, if there is a mine in that position. This probability gives indirectly a measure of the non-detection probability of that material as well.

- The *probability of false alarm* is the probability of having an alarm, for a given material, in a given location if there is no mine in that location.

The two latter definitions are extremely important to understand the humanitarian demining problem and for designing demining systems.
It is indeed particularly important that the detection probability should be as close as possible to one. It is easy to show that evaluating the detection probability also amounts to evaluating the risk of the occurrence of a mine which has not been detected. This risk is concerned with *human preservation* and is therefore of the utmost importance. No such risk is acceptable and it is therefore an absolute requirement that a demining system should decrease the probability of such a risk to the lowest upper bound possible (UN requires 0.4% maximum).
Besides, although one indirectly saves human lives by decreasing the false alarm risk thanks to the acceleration of the demining operations, the false alarm risk is also a question of cost. Indeed, a demining method which minimizes the false alarm rate results in an acceleration of the demining operations which results in spending less money.

Therefore, any demining operation enhancement must result in the highest possible detection probability (close to one) and in the smallest possible false alarm rate and that at the lowest price. Generally, it is accepted that the most efficient way for increasing the detection probability while minimizing the false alarm rate consists in using several complementary sensors in parallel and in fusing the information collected by these sensors.

As a matter of fact, it is imperative to evaluate the detection probability when optimizing the performances of a system. However, the detection probability, as it is defined before, assumes that a mine is present in the considered position. Since, during organized trials, the position of the mines is well known, the condition of the occurrence of a mine in the given position where the performances of a system must be evaluated is always realized. This latter remark is of particular importance because it justifies the organization of trials and the construction of models, to be validated by trials, in order to evaluate the detection probabilities (see also section 2.1).

Furthermore, assuming in the following as the first approximation that the sensors are independent[2], the detection probability can be maximized by optimizing separately the design of each sensor and of the associated signal processing. Next, it can easily be shown that the detection probability increases if the number of different sensors increases and that maximizing the overall detection probability of a set of independent sensors clearly comes to the same as maximizing the detection capabilities of each individual sensor. This justifies the use of several complementary sensors and of data fusion techniques to increase the detection probability. Among the most cited sensors one finds the metal detectors, the radars and the infrared sensors.

Finally, the false alarm risk, i.e. the probability of having an alarm if there is no mine, cannot be as easily evaluated as the detection probability because of the use of data fusion methods which favor the manual or automatic cancellation of false alarms. Furthermore, it is very difficult to evaluate the risk of false alarm because it is very difficult to define in a general way what is not a mine. In this context, it should be particularly inappropriate that a demining system, whatever it may be, makes decision instead of the final user whose own physical security is involved. Therefore, a

[2]the independence is a particular case of the complementarity: two independent sensors are complementary, but the contrary is not necessarily true.

well designed system should help the user in the decision making, not by replacing him, but by implementing efficient data fusion methods. For this purpose, methods which are able to deal with uncertainty by making proposals including the doubt to the user seem to be promising.

The Belgian humanitarian demining project HUDEM has been initiated by the Belgian Ministry of Defense and is supported by the Belgian Ministry of Defense and the Belgian State Secretariat for Development Aid. It is carried out in collaboration with laboratories of other Belgian universities, i.e. the "Facultés universitaires Notre-Dame de la Paix" (FUNDP), the "Katholieke Universiteit Leuven" (KUL), the "Universiteit Gent" (RUG), the "Université catholique de Louvain" (UCL), the "Université de Liège" (ULg), the "Université libre de Bruxelles (ULB), the "Universitaire Instelling Antwerpen" (UIA) and the "Vrije Universiteit Brussel" (VUB), and it is coordinated by the Royal Military Academy (RMA) [2]. This research project aims at contributing in solving the acute human problem of mine pollution by funding research grants devoted to basic research on mine detection and removal.

The first step of this paper consists in acquiring knowledge on sensors by means of trials and this is explained in section 2. As explained in section 3, the second step consists in developing models for the description of the ground electromagnetic behavior, in investigating the capabilities of new sensors (hyper-spectral imagery, nuclear quadrupole resonance, ... and educated rodents) and in enhancing the capabilities of existing sensors (ground-penetrating radars, metal detectors and infrared sensors). The third step means making each of these sensors skilled specialists of their respective domain (e.g. mine metallic content detection for the metal detector), as explained in the section 4.1 which analyses specific preprocessing tools and in section 4.2 which describes some dedicated pattern recognition tools. The last steps is sketched out in section 5 consist in fusing the high level information produced by the different experts (the sensors with their dedicated processing tools).

2 Outdoor trials

2.1 Multi-sensor data collection trials - The MsMs project

As has already been said, in the field of humanitarian demining more and more people feel that fundamental advances can be made by using a combination of mine detection sensors, as opposed to a single sensor.

In this multi-sensor approach, an important part of the research deals with the question of how to combine the (information coming from the) different sensors in an optimal way. A major problem however along this main research axis is the lack of (well-documented and well-registered) multi-sensor data. This kind of data is indeed needed in order to develop, experiment, and validate sensor or data fusion (to be more general) algorithms.

In order to cope with this lack of multi-sensor data, the Joint Research Centre (JRC) of the European Union has decided to sponsor a Joint Multi-sensor Mine-signatures measurement campaign (called the *MsMs project*), which is managed by the Royal Military Academy (RMA).

The objective of the MsMs project is to organize and execute an experimental campaign for collecting well documented and registered data of buried land-mines with multiple sensors. These data sets will then be made widely available to researchers and developers working on sensor fusion, signal processing for improved detection and identification of land-mines, assessing the role of the operator in the detection process, etc .... The outdoor test facility (5.7 x 80 m) of the Joint Research Facility of the European Commission, located at Ispra (Italy), houses the test minefield. Six test strips of 6 x 6 m consisting of different soil types (cluttered grassy terrain, loamy soil, sandy soil, clay soil, soil with high content of organic matter, and ferromagnetic soil) are complemented with one reference test strip of 6 x 6 m consisting of pure sand. The list of objects buried in the minefield includes mine simulants of three different dimensions with either a low or a high metal content, reference targets for position referencing and calibration checking, and clutter objects including empty bullet cartridges, metal cans, barbed wire, stones, wood, plastic boxes, etc .... This test minefield is going to be left intact for a long period, in order to be able to perform multiple runs on it. For the test campaign of the year 2000, the core sensors were a metal detector, a ground penetrating radar, a microwave radiometer, and thermal infrared imagers. Later, other (more experimental) detectors will also be tested on the same test minefield. The first data sets are in the process of being released right now [19].

2.2 Data collection in Belgium

In addition to the MsMs trials and JRC setup, realistic dummy minefields have been installed in the RMA and in MEERDAAL by the Belgian Bomb Disposal Unit (SEDEE/DOVO), the RMA and the Army engineering School. The latter test site is made of the four lanes (natural soil, sand, gravel and sand+gravel) where the mines and the false alarms are laid. Infrared images data were recorded every 15 minutes during 24 hours in two spectral bands (3-5 μm and 8-12 μm) parallel to the ground surface, to avoid interfering reflections. These data, collected on 5 CD-ROM's, are available on the JRC web-site (http://www.tdp.sai.jrc.it/APL-Database/Home/sigdata.htm). These trials were organized with the help of the technical service of the Belgian Army (DTT). Several Ground Penetrating Radar (GPR) experiments organized by the RMA (SIC) also took place, particularly with the RAMAC, the SIR2, the ERA GPR's[3] and the laboratory UWB GPR build by RMA. The results stored on CD-ROM are available on the same JRC web-site.
Recently, approximately 400 dummy AP mines were laid in Meerdaal. This new test site includes two minefields, the first one is design to tune existing or new equipments, the second one is design to perform blind tests. The two minefields, the configuration of which is described in Fig. 1, include different types of soil, work cells (where the de-miners can implement their system) and test cells (where the mines are).

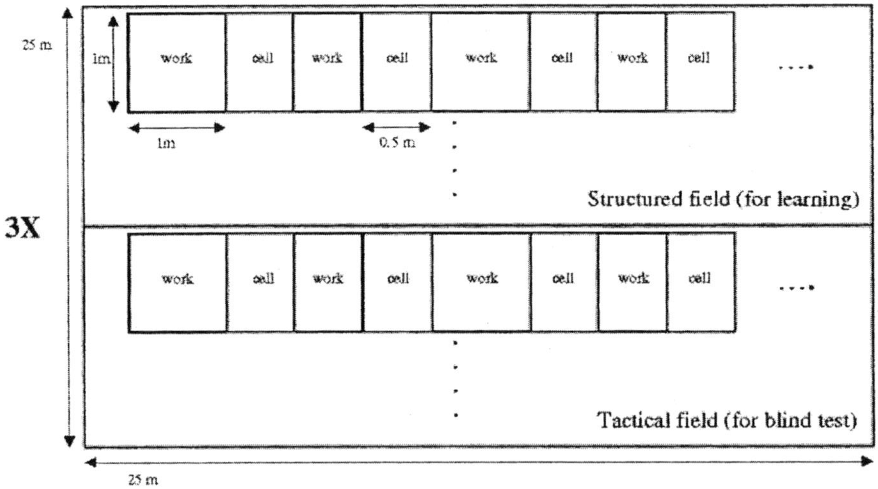

Figure 1: Meerdaal new minefields layout

[3]Respectively from Malå Geosciences (SW), Geophysical Survey Systems, Inc (GSSI) and ERA technology, UK.

3 Sensor design and analysis

3.1 Educated rodents

Actually, one of the most efficient "sensor" for mine detection is the dog. But it appears that rodents are easier to educate and to feed and that they can work longer than dogs. Furthermore, the rodents are much lighter and have a better olfactory capacity and a better immunity. The non profit organization APOPO and the University of Antwerpen are currently making tests in educating rodents and have already booked interesting results in Tanzania.

Rodent

3.2 Ultra-wide band (UWB) ground penetrating radar (GPR)

Useful definitions to understand what follows have to be given first. An A-scan is a one-dimensional signal taken perpendicular to the ground surface and is the basic echo signal produced by a GPR. A B-scan is a two-dimensional signal resulting from a collection of adjacent A-scans along a straight line horizontal to the ground surface. A C-scan is a two-dimensional horizontal slice (parallel to the ground surface) in a set of adjacent B-scans.

3.2.1 Soil characterization

Ground penetrating radars and passive radiometers are intended to function as anti-personnel mine detectors. Their performances depend upon parameters such as type and texture of soil, soil water content, soil density and operating frequency. In order to evaluate the performances of microwave technologies in land-mine detection, the electrical properties of soils have been extensively evaluated by the Microwaves Laboratory of the UCL from 2 to 18 GHz both theoretically and experimentally, using a new transmission method, as a function of soil type and water content. This has resulted in a new model for the microwave permittivity of soils, valid over a wide frequency range as shown in [15].

3.2.2 UWB antennas development

The Electrical Engineering and Telecommunication Departments of the RMA have developed small-sized ultra-wide band TEM horn silicon-filled antennas with a frequency range from 1 GHz to 7 GHz [4, 14, 3]. The maximum size of these antennas does not exceed 12 cm. By using of the shelf equipments (i.e. a pulse generator for the pulse UWB radar, a vector analyser for the UWB stepped frequency radar, a computer controlled digitizing oscilloscope and a computer controlled X-Y gantry on which the emitting and the receiving antennas are fixed), it has been possible to record, in both modes (pulse and stepped frequency modes), the A-, B-, and C-scans which are needed for our project and for optimizing the antennas. Fig. 2 shows the two antenna plates which were used for the tests and a sample A-scan of a PMN mine in loam. Fig. 3 presents two B-scans of a PMN mine respectively in loam and in sand and a C-scan of a PMN mine in loam at 4cm depth. The special design of the antennas allows the combined use of a GPR and a metal detector.

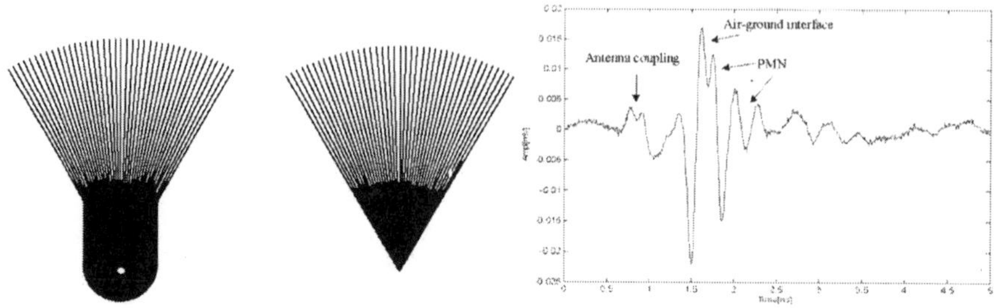

Figure 2: Antenna plates and an example of a PMN A-scan in loam

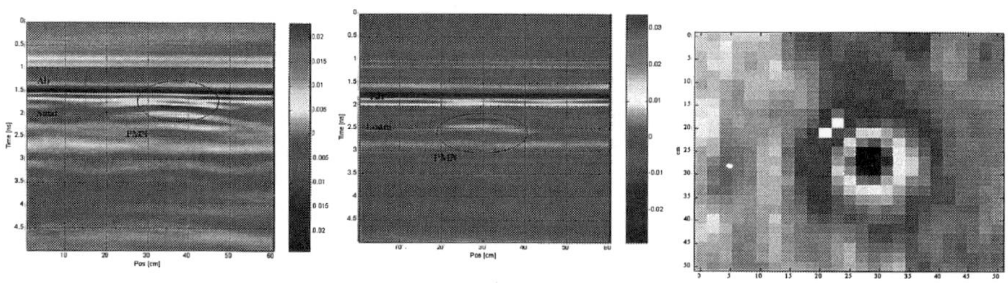

Figure 3: From left to right, PMN B-scans in sand and in loam and a PMN C-scan in loam

3.3 Modeling emitted IR radiations

Because it is impossible to control the IR environment (position of the sun, sky radiance, air temperature, background reflections and emissions, etc...), it has been decided to design a simple model that gives information on the temperature evolution on the surface of the ground above anti-personnel mines and its surroundings in order to be able to answer questions like "why, when, where and which thermal IR sensors to use ?". Only thermal conduction has been analyzed. The model is based on the general equation of heat conduction with specific assumptions (symmetry, simplified shapes, constant parameters, etc...). A numerical model, based on a finite volume method, has been developed in Excel and can be very easily modified for a set of different parameters such as mine shape, soil type and depth. This work was done by the ULB with the support of the RMA and the VUB [12, 11].

3.4 Hyper-spectral images

In the context of this project and of a collaboration between the Inter-university Microelectronics Centre (IMEC) and the RMA, taking into account the very selective properties of the material reflectivity, it has been possible to demonstrate the capabilities of wavelength tuning for discriminating different materials. Very narrow wavelength bands have been used in laboratory experiments. The results have shown that hyperspectral sensors could be very useful to discriminate shallow burried objects.

3.5 Nuclear quadrupole resonance

Advanced Nuclear Quadrupole Resonance (NQR) techniques can be used to detect explosives in any surroundings. The quadrupole charge distribution of the atom results in alignments of nuclear spins. A radio frequency pulse (RF-pulse) generated by a transmitter coil causes the excitation of nuclear spins to higher quantized energy levels. When the nuclear spins return to their equilibrium position, they follow a particular precession frequency. This specifies the atoms and functional groups in the molecules. Nitrogen is a quadrupole atom that appears in every type of explosive. Because of very distinct NQR frequencies the false alarm rate due to other nitrogen containing materials is extremely low [6]. The study of this type of sensor results from a collaboration between the King's College in London, the RMA and the KUL.

4 Signal and Image processing

The aim of this section is to show how to build skilled specialists from each raw sensor, using signal processing techniques. The first subsection 4.1 addresses the problem of the signal conditioning or preprocessing, *i.e.* signal detection, noise reduction, signal restoration and enhancement, which is a very important step before further processing. The second subsection 4.2 deals with the problem of recognizing the signal content by applying pattern recognition techniques aiming at increasing the expertise of each sensor separately.

4.1 Data preprocessing - noise reduction & restoration

4.1.1 GPR data preprocessing

In the case of GPR signals, A-scans and B-scans were studied. Useful signals, i.e. containing useful information, can be extracted and the signal to noise ratio of objects detected in A-scans or B-scans has been enhanced. In this context, the VUB has developed [18] a method for signal detection in A-scans using binary hypotheses test (signal is either background or useful), after having removed an averaged A-scan from the B-scan.
The RUG has been investigating [10] the applicability of multi-resolution decomposition techniques to remove the background of GPR images with buried mines. The main novelty is use of realistic statistical distributions of the wavelet coefficients which represent mainly speckle noise at one hand and those that represent the useful signal corrupted by speckle at the other. Analytic models have been derived for these distributions, and an automatic computation method of their parameters directly from a given image has been developed. The resulting algorithm greatly suppresses speckle, while preserving image details and its sharpness. The second image of Fig. 4 shows the result after horizontal filtering, which means simple subtraction of the average trace from each row. The third and the fourth images are the reconstructed version by the multi-resolution scheme with a non-separable filter and a separable filter respectively.

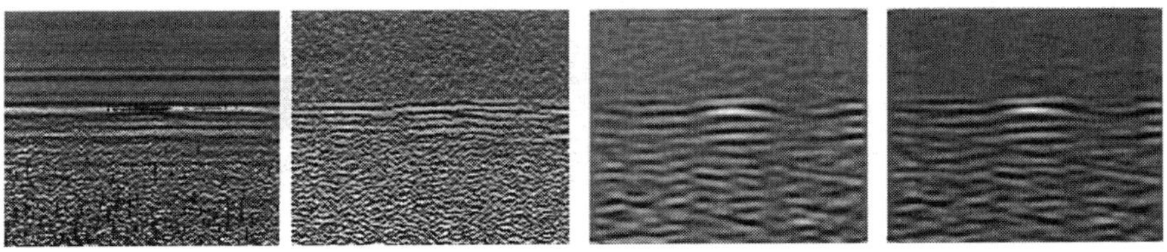

Figure 4: GPR B-scan - background and speckel noise removal experiments

A principal component analysis of a sequence of C-scans has been done by the VUB [17] and produces also interesting results shown on Fig. 5: sample C-scan images of the sequence on the left side and the four most significant images (with the largest variances) after the Karhunen-Loève transform on the right side.

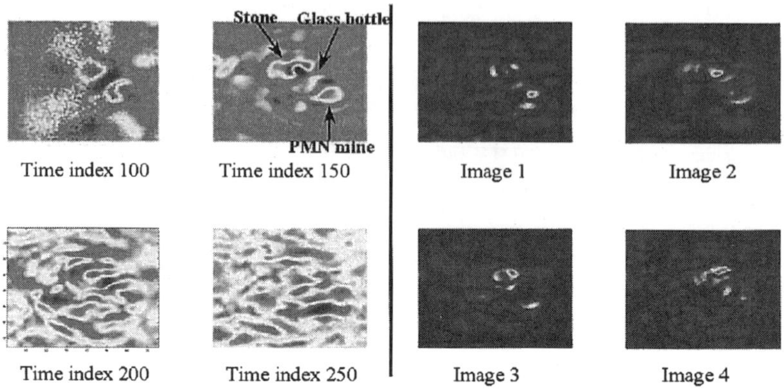

Figure 5: GPR: Principal component analysis of a C-scan sequence

The processing of the absolute value of A-scans using the Hilbert transform has been done by the RMA in order to obtain the A-scan envelops and to enhance the resulting C-scans. The results obtained with this method are shown on Fig. 6.

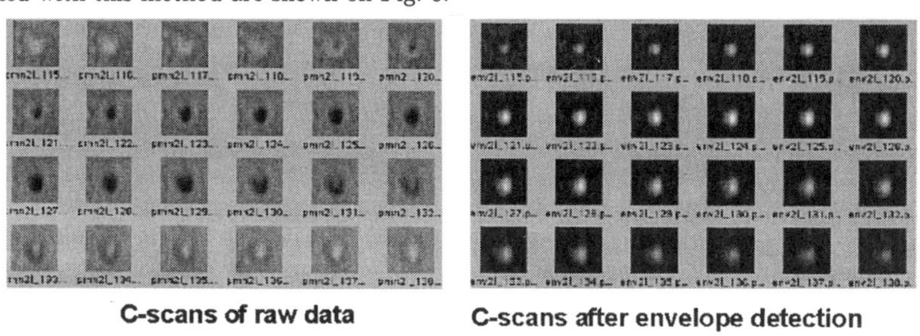

Figure 6: GPR: enhancement of a C-scan sequence using the Hilbert transform

RMA [13] has decomposed A-scans in a linear combination of wavelets. A limited number of wavelet coefficients (typically 5 coefficients) is sufficient to represent A-scans with a good resolution. The selected mother wavelet is derived from the emitted radar impulse. The results are shown on Fig.7.

4.1.2 IR data preprocessing

Efficient de-noising methods in the wavelet domain have been proposed by the RUG [9]. A first method adds spatial constraints to the criterion for selecting noisy wavelet coefficients and for each coefficient the probability of being noise-free is computed. The spatial constraints are derived from prior geometrical assumptions expressing the fact that meaningful wavelet coefficients

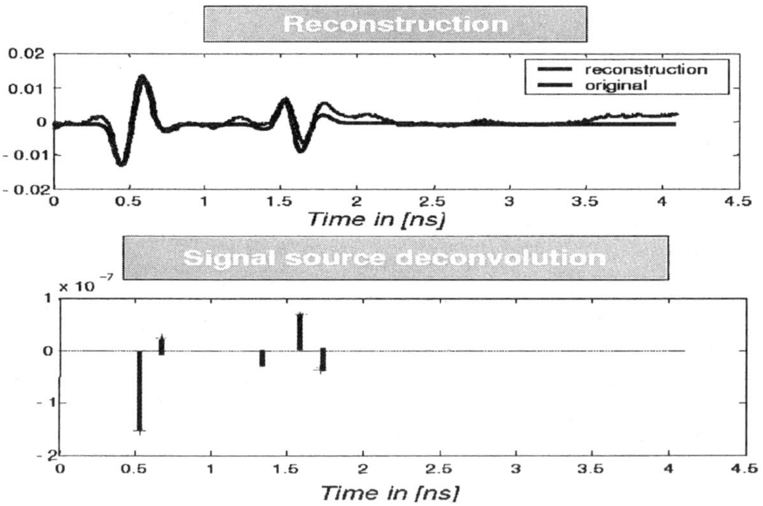

Figure 7: GPR: representation of A-scans with a limited number of wavelet coefficients

appear in spatially connected clusters, at the location of characteristic image features like edges, corners, etc.... For the criterion itself the magnitudes of the wavelet coefficients are used. Fig. 8 summarizes the results. In a second method, a wavelet decomposition in three resolution levels,

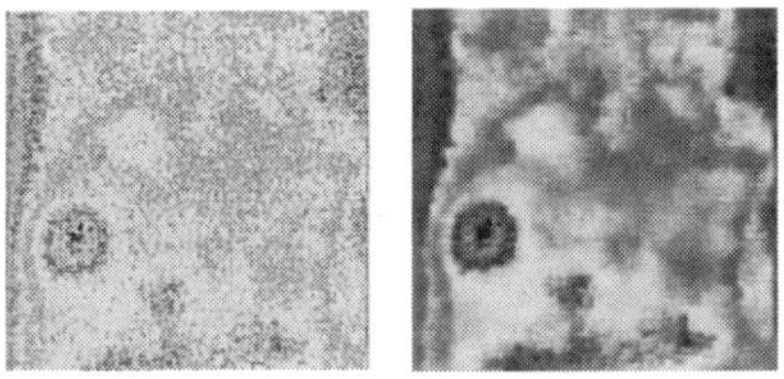

Figure 8: Original IR images and corresponding restored IR image

with one approximation and one detail image at each scale, is applied to the image. A binary mask, obtained by thresholding the approximation images on which an edge detection algorithm has been preliminary applied, is used to preserve the edges during a non linear filtering operation on the wavelet coefficients before the reconstruction phase. Fig. 9 summarizes the results.

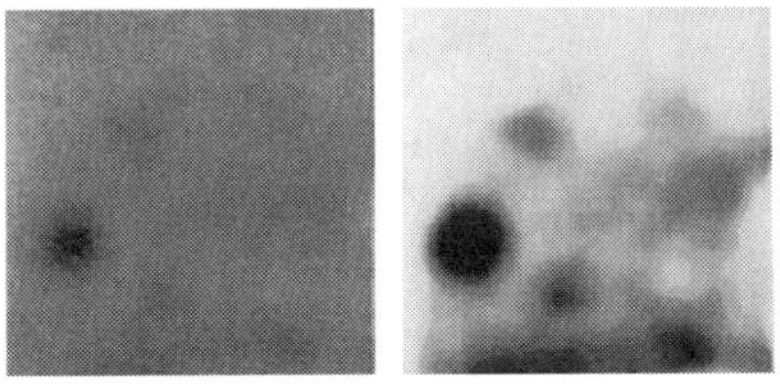

Figure 9: Original IR images and corresponding restored IR image

The analysis of an IR sequence by means of the Karhunen-Loève transform or the Kittler-Young transform (the latter however requires a learning phase, which is obviously not recommended in humanitarian demining applications because of the high diversity of possible scenarios), leads to interesting results [12] as well. Fig. 10 shows the obtained results. On the left side, sample IR images of the sequence are presented. On the right side, the two most significant images (with the largest variances) after transformation are presented for the two transforms.

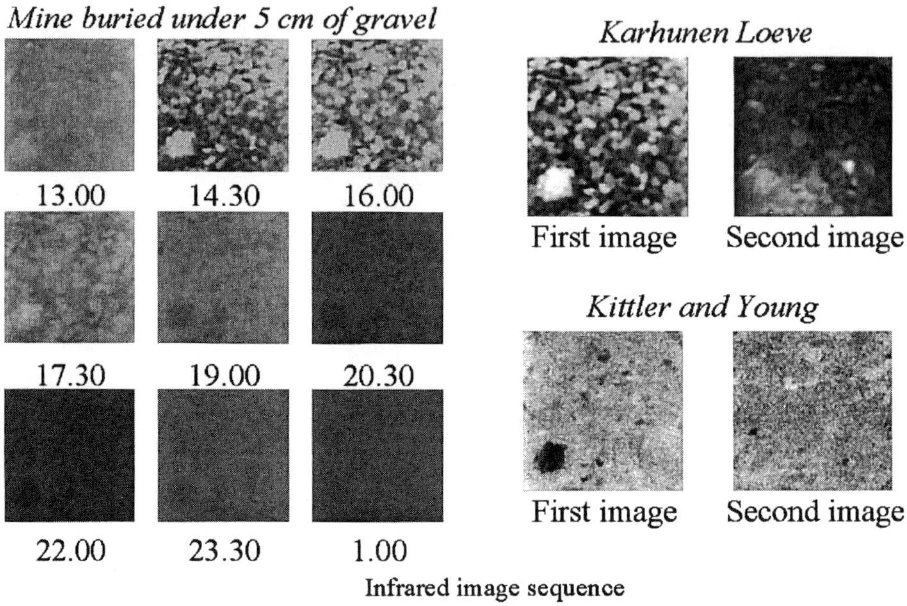

Figure 10: Karhunen-Loève and Kittler-Young transforms

4.1.3 Metal detector image preprocessing

Surprisingly, the metal detector (which is the most common detector), considered as an imaging device, can also provide very useful information on the shape of metallic pieces included in mines. Trials have been performed in the RMA [16] by placing the Vallon metal detector on a gantry. Thanks to the collaboration of Vallon, the metal detector has been provided with a digital output. The obtained images were processed as follows: after noise reduction using a Wiener filter, the images were de-convolved using respectively an experimental point spread function (PSF), which was recorded using a small metallic ball, and a simplified theoretical model of the PSF, calculated from the detector head shape. A practical example, using the exercise mine PRB409 which has a diameter of 7.5 cm, is shown on Fig. 11. The image on the left is the original one. The central image is a RX photography of this mine where one sees clearly a small horizontal metallic cylinder on the right side. The image on the right shows the de-convolved image where the metallic cylinder clearly appears.

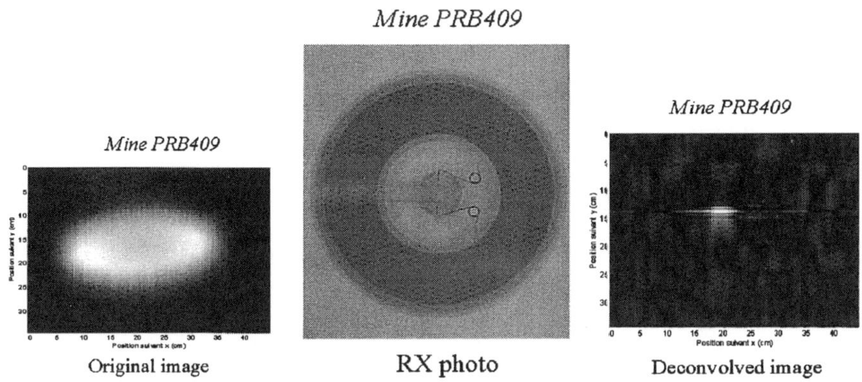

Figure 11: Metal detector — PRB mine de-convolution

Unfortunately, the PSF of a metal detector is a function of the depth (see Fig. 12) and of the nature of the buried metallic object (eddy currents are different in a close and in an open circuit) and the image formation process is non linear. However, the in depth modeling of the metal detector behavior as a function of the type of buried object by the RMA [5], has shown that it is possible to derive the depth of a buried object from the original data and thus to derive the corresponding PSF to allow a correct de-convolution (see Fig. 13, 14 and 15). Further, information on the symmetry properties of the burried metallic objects can easily be extracted. This subject is still under investigation. This interesting consideration shows that the metal detector, known as a cheap mine detection system, remains a promising device.

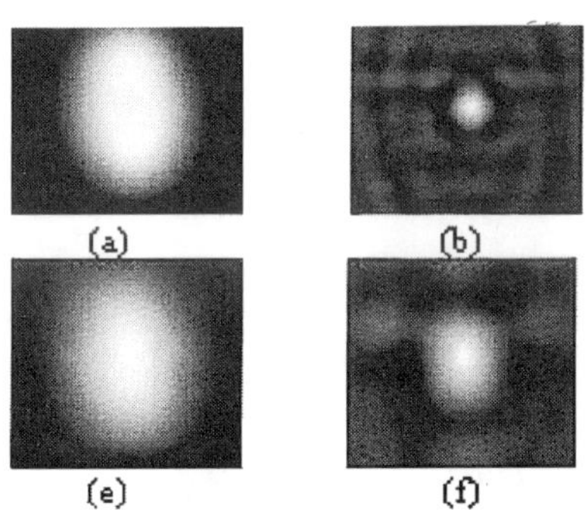

Figure 12: Metal detector — deconvolution of a metallic ball at 5.5cm and 8.0 cm

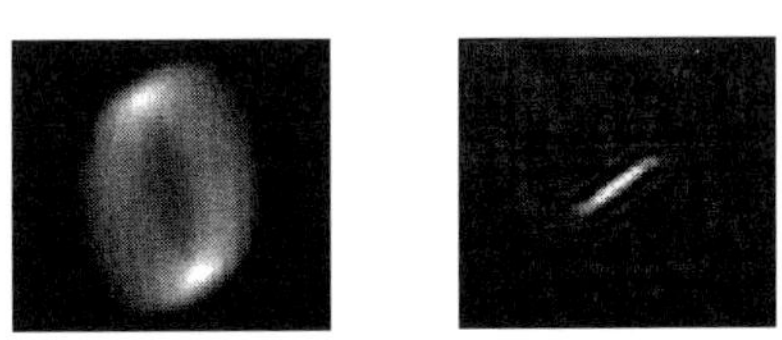

Figure 13: Metal detector — deconvolution of a straight metallic wire

Figure 14: Metal detector — deconvolution of a metallic open loop

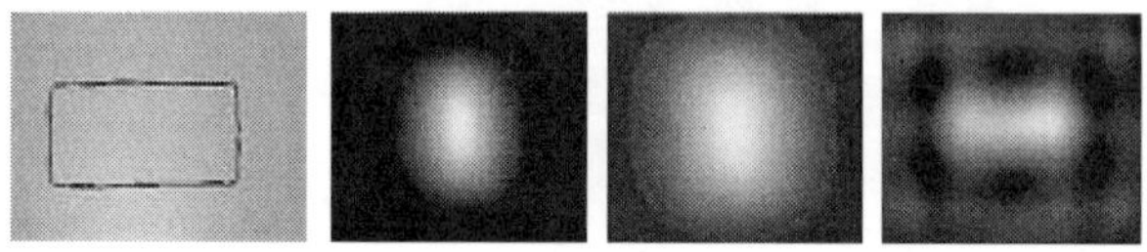

Figure 15: Metal detector — deconvolution of a metallic closed loop

4.2 Mine detection expert development

After the signal conditioning, the next step consists of applying pattern recognition methods to the preprocessed data of each sensor in order to get a semantic description of the data information content.

4.2.1 GPR expert

The approach used by the RMA for all three experts (MD, GPR and IR) is to extract information about shape of objects from their raw data [7].

In the case of GPR, on each C-slice simple preprocessing (including edge detection) is performed and important edges are extracted. The edges from all C-slices are then put together either on the same 2D image (on Fig. 16[4] edges from the same depth, i.e. same C-slice, are presented by the same color) or on a 3D image, in which each C-slice represents a certain depth. If the A-scan

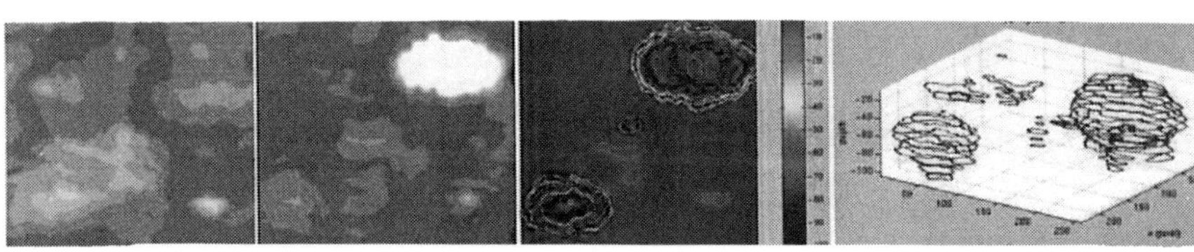

Figure 16: From left to right, two C-slices followed by a 2-D image of edges from all C-slices, where color is a label of depth, followed by a 3-D image of edges

envelops are detected using the Hilbert transform, as previously explained, the results are much better (RMA) as it can be seen on Fig.17.

In order to recover the correct 3-D shape of buried objects, RMA [13] has developed algorithms based on the convolution, by modeling the behavior of the GPR in the time domain. The developed algorithms are faster than the classical migration methods and provide very good results as it can be seen on Fig.18.

[4]Data provided by ERA Technology, UK

Figure 17: From left to right: view of a mine, 3-D GPR image of the same mine

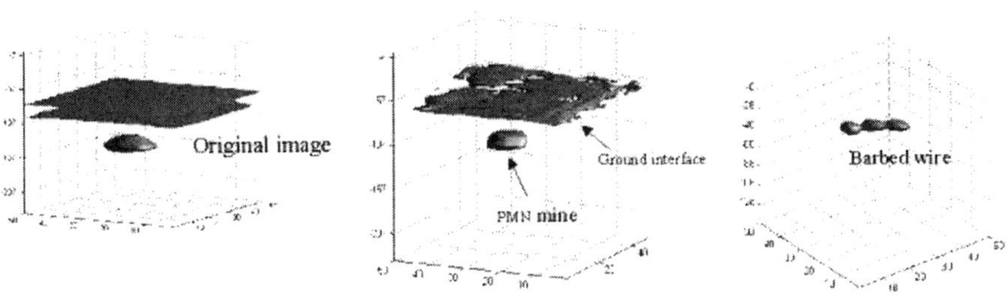

Figure 18: From left to right: raw GPR 3-D image of a PMN mine, restored 3-D image of a PMN mine and restored 3-D image of a 10cm barbed wire with three barbes

The VUB has developed a classification method for ultra-sonic and GPR signals [18]. The method first consists in extracting features (the time signal itself, its Fourier transform, its auto-correlation, its wavelet coefficients, its Wigner-Ville transform and the derived scattergrams of the latter). After a selection of the most discriminant features, a supervised parametric approach based on the Bayes optimal classifier (which requires a training phase, which is again not recommended in humanitarian demining applications because of the high diversity of possible scenarios), is used. A better approach based on the Dempster-Shafer framework is presented in section 5. Two classification methods have been tried: the first one only implements one classifier making use of all the selected features, the second one uses a hierarchically organized multi-classifier system, combining the conditional probabilities computed by specialized classifiers either by averaging them or by multiplying them. The obtained results are described on Fig. 19. The GPR data of these tests were provided by the DETEC laboratory of the "Ecole polytechnique fédérale de Lausanne" (Switzerland).

4.2.2 IR expert

For images taken by infrared cameras, the RMA has performed a simple preprocessing, followed by edge detection. Since most of the mines are circular, they will be ellipses on binary, edge detected images. Because of occlusion and noise, these ellipses are generally not continuous. Therefore, the RMA developed shape recognition tools that are able to detect partially occluded ellipses [7]. Two examples of obtained results are shown on Fig. 20.

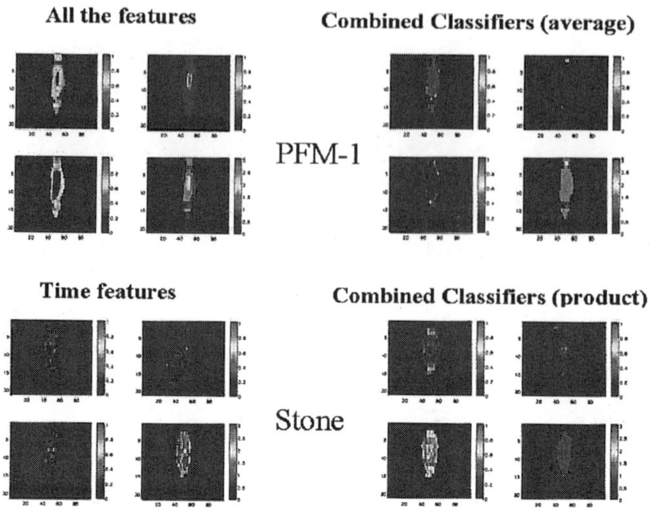

Figure 19: GPR: Classification of a stone and of a PFM-1 mine in the ground from C-scans

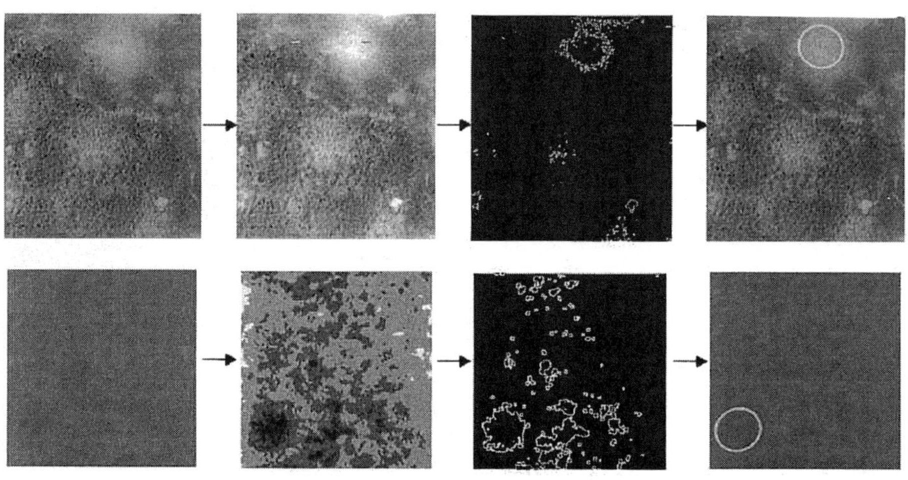

Figure 20: From left to right: the original image followed by the preprocessed image, the detected edges and the original image with detected mine-like shape (ellipse)

4.2.3 Metal detector expert

Taking into account that there are also some mines that do not have regular shape, the RMA is analyzing possibilities of recognizing arbitrary shapes [7]. Furthermore, in the case of plastic mines, just a small metallic part inside such a mine (*e.g.* fuse) will be detected, and that shape does not have to be elliptical. An example is given on Fig. 21 below, where the detection of an X shape is performed.

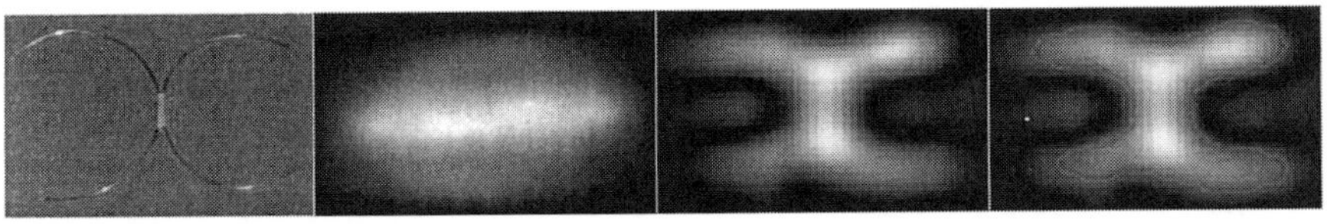

Figure 21: From left to right, the used X shape, the output of the metal detector, the de-convolved output and the detected shape

5 Expert fusion

The analysis of the principles of operation of an imaging metal detector, a GPR and an infrared camera, their complementary information, and the factors that affect their operability, have led us to the conclusion that their fusion should result in improved detectability and reduced number of false alarms in various situations (different types of mines, of soil, vegetation, moisture, etc...). Therefore, we have been analyzing possibilities to combine these three sensors, but the model we have been developing is quite general, *i.e.* it can be easily modified to include other sensors as well. Since in this domain of application we have to deal with uncertainty, ambiguity, partial knowledge and ignorance, we have chosen for an approach where they can be appropriately modeled: belief functions within the framework of the Dempster-Shafer theory. A main motivation for working within this framework is to be able to easily model and include existing knowledge regarding: chosen mine detection sensors, mine laying principles, mines, and objects that can be confused with mines.
In order to illustrate our approach, let us analyze a very frequent case in reality: the detection of high-metal content objects. Since all three sensors give images, *i.e.* information about the size of an object, the following four classes create the frame of discernment Θ:

- MMR (metallic mine of regular shape),
- MMI (metallic mine of irregular shape),
- MFR (metallic friendly, *i.e.* non-dangerous, object of regular shape),
- MFI (metallic friend of irregular shape).

Furthermore, the criteria that can give the most information about the real identity of the object in this case, for our knowledge, are the following:

- For each of the three sensors:
  1. Ellipse fitting (that is, how well the shape of the object fits in an ellipse), assigning masses to subsets {MR, FR}, {MI, FI}, Θ;
  2. Shape elongation, again giving masses to {MR, FR}, {MI, FI}, Θ;
  3. Area/size (by which information about expectable size range of mines is included), assigning mass mainly to Θ within that range, and to {FR, FI} elsewhere;
- For MD: burial depth (including the knowledge about the depths where mines can be expected), assigning masses to Θ and {FR, FI};
- For GPR:
  1. Depth dimension of the object, assigning masses to {FR, FI} and to Θ;

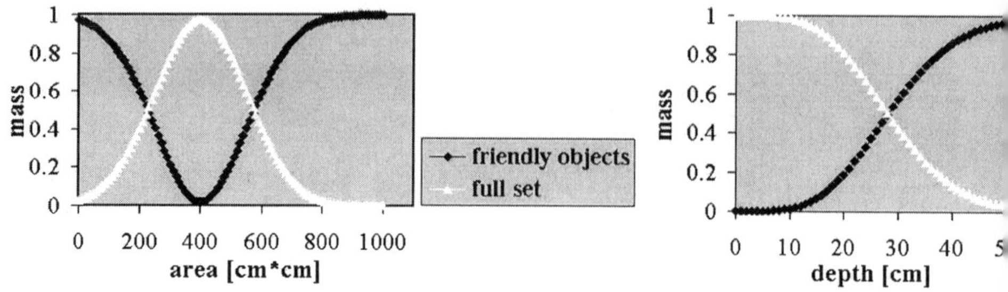

Figure 22: Mass assignments for the area (left) and burial depth (right) criterion

2. Comparison of the depth position of metal detected by MD and the object depth interval sensed by GPR: if they are in accordance, masses are assigned mainly to Θ, if they are not, the largest part of the masses should go to the subset {FR, FI}.

The masses are defined as functions depending on such measures as ellipticity, elongation factor, area, and depth. They are detailed in [8].
After assigning masses for all sensors and all criteria, we combine them using the well-known Dempster's rule in *unnormalized* form, in order to preserve potential conflict between the sensors. After that, guesses about the real identity of an object under observation are made, on the basis of the strength of belief assigned to each subclass of the frame of discernment. This list of guesses is served to a de-miner, together with confidence degrees, as well as processed data of each sensor, in order to help *him* in making his final decision.

6 Conclusions

The purpose of this paper is to present the sensor design and signal processing aspects of the Belgian HUDEM project, by describing the efforts and their results for increasing the knowledge on sensors and on sensor/ground characteristics, for designing new sensors or tuning old ones and for processing the data produced by sensors. Furthermore, this paper presents the detection as a global process wherein the outputs of the sensors, considered as skilled specialists, can be integrated in a fusion operation. The future work includes the further development of the sensor investigation and design, the enhancement of the signal preprocessing methods, the improvement of the dedicated expertise of each sensor and the development of associated effective data fusion methods. This implies as well the analysis and the exploitation of the results of the multi-sensor trials, conducted in the framework of the MsMs project.

Acknowledgments

The authors wish to thank all the researchers of the HUDEM and MsMs projects without whom it was impossible to write this paper. In spite of the difficult current international situation, it was possible for our foreign researchers to find the necessary resources for providing an excellent work and for maintaining an enthusiastic spirit of collaboration. They think particularly to N. Milisavljevic, A. Pizurica and D. Milojevic. They also wish to emphasize the excellent work performed by the Belgian researchers, A. Engelbeen, M. Storme, B. Scheers, L. van Kempen and B. Weetjens. They also wish to thank Prof I. Lemahieu, Prof. W. Philips, Prof. P. Van Ham, Prof.

J.A.S. Smith, Prof. Van Hecke, Prof. A. Vander Vorst, Prof. J. Cornelis, and Dr. A. Sieber for the support they provide to the researchers.

References

[1] M. Acheroy. L'infrarouge thermique, principes et applications au déminage humanitaire. *Revue HF*, (3):p. 13–24, 1998.

[2] M. Acheroy, Y. Baudoin, and M. Piette. Belgian project on humanitarian demining (HUDEM). In *Proceedings CLAWAR '98*, pages p.215–218, Brussels, Belgium, 1998. BSMEE.

[3] S. Acheroy. Proefondervindelijke studie van een breedbandige hoornantenne voor de detectie van ondergrondse mijnen. Master's thesis, RMA, Brussels, Belgium, 1998.

[4] M. Piette B. Scheers. Short-pulse response of anti-personnel landmines to UWB GPR signals. In *Seventh international conference on ground-penetrating radar*, Kansas, USA, 1998.

[5] P. Druyts, L. Merlat, and M. Acheroy. Modeling Considerations for Imaging with a Standard Metal Detector. In *Detection and Remediation Technologies for Mines and Minelike Targets V*, volume 4038, pages 1431–1451, Orlando, FL, USA, April 2000. SPIE Press.

[6] A. Engelbeen. Nuclear quadrupole resonance mine detection. In *CLAWAR'98*, pages pp.249–253, Brussels, Belgium, November 1998. BSMEE.

[7] N. Milisavljević. Comparison of Three Methods for Shape Recognition in the Case of Mine Detection. *Pattern Rcognition Letters*, 20(11-13):1079–1083, 1999.

[8] N. Milisavljević, I. Bloch, and M. Acheroy. A first step towards modeling and combining mine detection sensors within Dempster-Shafer framework. In *The 2000 International Conference on Artificial Intelligence (IC-AI'2000)*, Las Vegas, USA, 2000.

[9] A. Pizurica. Multiresolution techniques for image restoration in mine detection problems. In *CLAWAR'98*, pages pp.225–230, Brussels, Belgium, November 1998. BSMEE.

[10] A. Pizurica, W. Philips, I. Lemahieu, and M. Acheroy. Speckle Noise Reduction in GPR Images. In *International Symposium on Pattern Recognition "In Memoriam Prof Pierre Devijver"*, Brussels, Belgium, Februari 1999. Royal Military Academy, RMA.

[11] M. Schachne. Détection de mines antipersonnel par sollicitations thermiques dans différents types de sols. Master's thesis, ULB, Brussels,Belgium, 1998.

[12] M. Schachne, L. van Kempen, D. Milojević, H. Sahli, P. Van Ham, M. Acheroy, and J. Cornelis. Mine detection by means of dynamic thermography: simulation and experiments. In *CLAWAR'98*, pages pp.255–259, Brussels, Belgium, November 1998. BSMEE.

[13] B. Scheers, M. Piette, M. Acheroy, and A. Vander Vorst. A laboratory UWB GPR system for landmine detection. In *GPR2000*, Sidney, Australia, June 2000.

[14] B. Scheers, M. Piette, and A. Vander Vorst. The detection of AP mines using UWB GPR. In *IEE Second International Conference, The Detection of Abandoned Land Mines*, Edinburgh, UK, October 1998. IEE.

[15] M. Storme, I. Huynen, and A. Vander Vorst. Characterization of wet soils from 2 to 18 GHz, experimental results. In *CLAWAR'98*, pages pp.237–239, Brussels, Belgium, November 1998. BSMEE.

[16] O. Thonnard. Détecteur de métaux - Analyse d'un détecteur de métaux et conception d'un système de balayage afin de produire des images 2-D du sol. Master's thesis, RMA, Brussels, Belgium, 1998.

[17] L. van Kempen. Signal processing and pattern classification techniques for AP mine detection. In *CLAWAR'98*, pages pp.231–236, Brussels, Belgium, November 1998. BSMEE.

[18] L. van Kempen, H. Sahli, E. Nyssen, and J.Cornelis. Signal processing and pattern recognition methods for radar AP mine detection and identification. In *IEE 2nd international conf. on the detection of abandoned landmines*, Edinburgh, UK, October 1998. IEE.

[19] P. Verlinde, M. Acheroy, G. Nesti, and A. Sieber. First Results of the Joint Multi-Sensor Mine-Signatures Measurement Campaign (MsMs Project). In *Detection and Remediation Technologies for Mines and Minelike Targets VI*, volume 4394, Orlando, FL, USA, April 2001. SPIE Press.

Electro-Optical Land Mine Detection

Ingo Schwaetzer
Bodenseewerk Gerätetechnik GmbH
88641 Überlingen, Germany

Abstract

Contracted by the German Ministry of Defense, BGT (Bodenseewerk Gerätetechnik GmbH, Überlingen) is responsible for the electro-optical sensor systems and the associated real-time signal processing within the (only) two German programs related to land mine detection. These two programs are MMSR (Mobiles Such- und Räumgerät, Mobile Ground Based Land Mine Detection and Clearing System) and AAMIS (Aufklärungs-Ausstattung Minensperren, Airborne Autonomous Land Mine Field Detection System.

The presentation covers four topics:

1. Basics on Electro-optical Land Mine Detection. Usable types of sensors. Detectable land mine features. Primary and secondary features. Examples of land mine images within the different spectra. Performance limits of electro-optical methods.
2. Real-time signal processing. Data rates. HW architecture. SW architecture. Data fusion from different spectra.
3. Description of AAMIS. Current trial results. Photographs of equipment and scenarios of current trials. Way to proceed and status of the program in respect to fielding the system.
4. Description of the electro-optical part of MMSR. Current trial results. Photographs of equipment and scenarios of current trials. Way to proceed and status of the program in respect to fielding the system.

H. Schubert and A. Kuznetsov (eds.), Detection of Explosives and Landmines, 57.

DECISION-TAKING PROCEDURE FOR EXPLOSIVES DETECTION BY NUCLEAR TECHNIQUE

D.N. VAKHTIN, A.V. EVSENIN, A.V. KUZNETSOV, O.I. OSETROV, M.D. ZUBKOV
V.G.Khlopin Radium Institute
194021, 2nd Murinsky pr., 28, St.-Petersburg, Russia

1. Introduction

One of the most promising techniques of explosives detection is based on detection of secondary γ-rays that are induced in the inspected object or area by a flux of neutrons (the so-called "neutron-in gamma-out" method). Fast neutrons undergo mostly inelastic scattering (n,n'γ) while thermal neutrons undergo mostly radiation capture (n,γ) reactions on nuclei inside the object. Spectra of secondary γ-rays contain information about the isotopic composition of the object. By using narrow (few nanoseconds) time gates one can determine the time of arrival of the given γ-quantum with respect to a particle which accompanies neutron emission from the neutron source, and thus split the total γ-spectrum into fast (inelastic scattering) and slow (radiation capture) components.

In our case, the inspected object or area is irradiated by neutrons from the isotopic source 252-californium placed inside an ionization chamber which detects the fission fragment that accompanies neutron emission from the source. The intensity of the source is about 4×10^6 neutrons per second, and the efficiency of fission fragment detection is >95%. Secondary γ-rays are detected by a γ-detector based on ∅150mm×100mm NaI(Tl) crystal.

Apart from secondary γ-rays from the inspected object, the measured γ-spectra contain contributions from cosmic and other uncorrelated background, as well as γ-rays from reactions of neutrons on construction materials and surrounding objects (correlated background). By placing nanosecond time gates on energy spectra we were able to reduce the correlated background component in fast (inelastic) spectra by an order of magnitude, and the uncorrelated background component by two orders of magnitude.

The major challenge is to extract the information on the inspected object from measured γ-spectra, and to be able to determine whether the object belongs to a class of explosives or to non-explosive objects, i.e. to "recognize" the object.

2. General approach to recognition

In order to recognize the object by its spectrum (in our case, by the spectrum of secondary γ-rays) one must first select certain finite number of characteristics of this spectrum that are different for different classes of objects, say, explosives and non-explosives. These characteristics can be regarded as coordinates in some multi-dimensional space, in which every individual measurement is represented by a single

H. Schubert and A. Kuznetsov (eds.), Detection of Explosives and Landmines, 59–67.

point. Since, due to a statistical nature of nuclear reactions, two different measurements would never produce exactly the same result, one has to deal with "clusters" of points, each cluster corresponding to responses of the experimental device to a certain substance. If one pre-measures (or simulates) spectra from some most common substances, one can define a number of such "clusters", each corresponding to a certain "hypothesis". For example, by measuring response of the experimental device to an explosive device, stones of different sizes, tree roots, pieces of metal, etc., one can determine the location of the corresponding clusters of points in the multidimensional space for each of these hypotheses.

The recognition procedure in our case then goes as follows: one measures the γ-spectrum, analyses it to extract its characteristics, and compares there characteristics with those of the known objects by determining the distance in multi-dimensional space from the single point corresponding to this individual measurement to clusters corresponding to different known objects. By comparing distances from the experimental point to different clusters one can calculate relative probabilities that the unknown object belongs to different kinds of objects. If the probability of belonging to a certain class exceeds some threshold value (say, 99.9%), then the object is recognized. If probability does not reach the threshold value, even when additional measurement with the same object are made, then none of the pre-defined hypotheses is true, and the object is unknown.

There are many ways to build the characteristics of γ-spectra for use in the recognition procedure, each method having its own strong and weak sides (see, e.g. [Vapnik (1995)]). Example of application of two such methods – least-square fitting (LSF) and principal component analysis with Mahalonobis distances (PCA/MD) – to the case of our experimental technique is described in the following sections.

3. Determining relative concentrations of chemical elements by the least-square fitting procedure (LSF)

In the LSF procedure the spectral characteristics that serve as coordinates in the multi-dimensional space for the recognition procedure are relative concentrations of several chemical elements, which can be found in a typical explosive. Almost any explosive contains carbon, oxygen, hydrogen, and nitrogen. Most of them also have some metal coating.

The natural coordinates are, thus, ratios of concentrations of oxygen, hydrogen, and nitrogen (C, H, O, N) to the concentration of carbon – n_O/n_C, n_H/n_C, n_N/n_C,. If there is no carbon in the measured spectrum, then it certainly belongs to a non-organic object, which cannot be an explosive.

In order to extract the relative concentrations of C, H, O, N from the measured spectra, one must know the background spectrum (which can be measured somewhere close to the inspected area), and the response functions of the experimental device to characteristic γ-lines from different elements. As a result of LSF one gets weights for each response function including the background, which ensure the best approximation of the experimental spectrum. These weights are related to concentrations of the corresponding isotopes in the inspected area.

We have developed an LSF procedure, which allows us to extract the concentrations of C, O, and N isotopes in the inspected object. Apart from C, O, and N, some other chemical elements are used in the fitting procedure: hydrogen, iron, chlorine, and silicon. Response functions of our experimental device to all these elements were calculated with a Monte-Carlo code, which takes into account geometry of the setup and the energy-dependent resolution of the γ-detector.

If the LSF procedure indicates that concentration of carbon n_C in lower than a certain threshold value, then the object in regarded as a non-organic substance, and thus recognized as non-explosive. Otherwise, ratios n_O/n_C and n_N/n_C are determined. The resulting distributions of n_O/n_C and n_N/n_C for 100 simulated spectra of 500g of TNT and 500g of tree root in open air are shown on Figure 1.

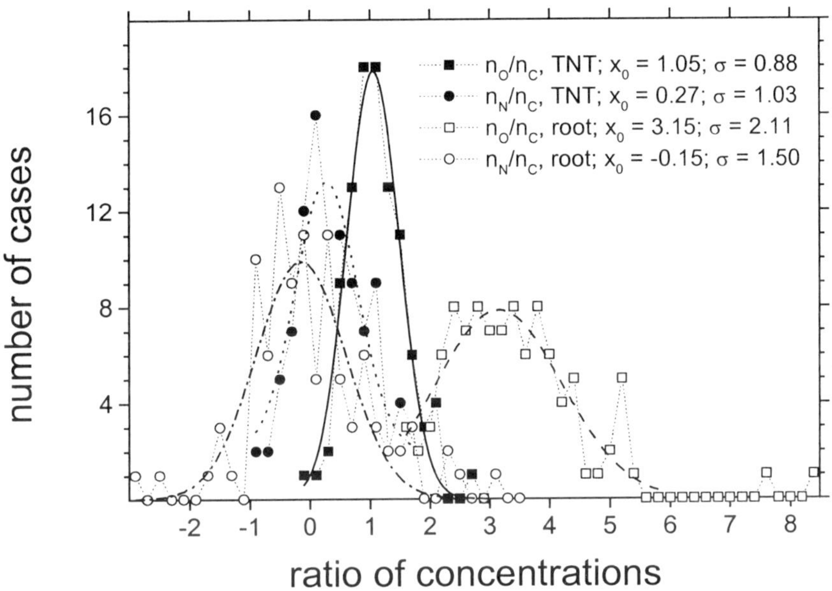

Figure 1. Distributions of ratios of concentrations n_O/n_C and n_N/n_C for 100 simulated spectra of 500g TNT and 100 simulated spectra of 500g tree root in air. Measurement time was 5 minutes. Legend on the plot shows the parameters x_0 and σ of Gaussian fits to the corresponding distributions.

One can conclude from the Figure 1 that the n_O/n_C ratio is a much more reliable "coordinate" for separation of TNT from tree roots. In order to reduce the intersection area of two gaussians, one must measure for a longer time. Another possibility is to use the n_H/n_C (hydrogen-to-carbon) ratio, but one must keep in mind that the number of 2.23 MeV γ-rays from reactions of hydrogen with thermal neutrons depends strongly on the number of neutrons that are thermalized in the medium (soil, sand, etc.), which surrounds the land mine. It is not clear how one can take into account the thermalizing properties of the soil, which is mostly dependent on its humidity.

The above parameters of gaussian distributions can be used to construct clusters in a two-dimensional space with coordinates n_O/n_C and n_N/n_C. Any individual measurement should be then analyzed to determine its location in this space. Distances from this point to centers of different clusters in the units of corresponding σ-s can be calculated, and then transformed into a relative probability that the spectrum belongs to either TNT or tree root.

4. The principal-component analysis with Mahalonobis distances (PCA/MD) method

4.1. IDEA OF THE METHOD

Another method of building a multi-dimensional space for recognition procedure is the principal component analysis with Mahalonobis distances (PCA/MD) [Cooley(1971), Gnanadesikan(1977), Sharaf(1986)].

The PCA method compresses the spectral data and decomposes spectra from the training set into common variations characterized by an orthogonal set of eigenvectors. From the procedure of decomposition into orthogonal eigenvectors one gets a small set of well determined "weights", which show a degree of presence of given variations in a spectrum. As the results of numerical experiments have shown, decomposition of the training set of response functions (spectra) from 10 samples into 2-4 components is optimal for the task of detecting explosive substances in presence of high background.

The method of Mahalonobis distances [3] allows one to determine the "distance" between an unknown spectrum and an average over the training set in terms of standard deviations obtained from the analysis of borders of the training set in the space defined by eigenvectors. The main advantage of this method is that it accounts for both common variations of spectra at single energy and correlations between different energy points of spectra (i.e. accounts for co-variations). In other words, practically all the information contained in the spectrum is taken into account. The method allows one to make reliable probability calculations, including cases of multidimensional analysis. Due to inclusion of certain number the first orthogonal component in the analysis, is possible to effectively filter statistical noise (contained, as a rule, in less important higher-order components).

In short, the main differences between using LSF and PCA/MD to extract "coordinates" of the multi-dimensional space for recognition procedure are that:

- eigenvectors in PSA/MD form an orthogonal basis, which means that all the clusters corresponding to different substances in a multi-dimensional space can be made spherical, while in LSF shapes of clusters depend on correlations between concentrations of different chemical elements;
- PCA/MD takes into account all the information contained in spectra, while LSF typically requires the selection of "interesting" and "not interesting" parts of spectra;
- in PCA/MD the number of orthogonal coordinates is determined by a general physical idea of experimental resolution and statistical noise (high-frequency noise is isolated in higher-order components and is not taken into account), while in LSF the number of coordinates (e.g., n_x/n_C) is based on some assumptions about the possible elemental composition of the inspected area.

4.2. PCA/MD ALGORYTHM

The above technique of the analysis of spectra adapted by us to tasks of explosives detection in conditions of bad statistics, includes the following basic stages:

1. Measurement of the background. It is necessary to create a library of 7-10 measurements of the background.

2. It is necessary also to create library of spectra of all those substances, whose presence is necessary to determine. It can be either experimental measurements with

good statistics, or calculated spectra obtained by the computer program that simulates response function of the experimental device.

3. For each group of substances whose presence we wish to determine, a library (training set) of spectra is created. Each spectrum of the library is a sum of the background and the reference spectrum of response to given substance. Depending on the threshold (minimal weight of the substance), the spectrum of the background and reference spectrum "mix up" in different proportions.

4. Preliminary processing of the spectral data from the training set (subtraction of an average background, normalization, smoothing etc.)

5. The spectra from the training set are decomposed into orthogonal eigenvectors (common variations). To obtain the basis of eigenvectors and appropriate weights we are using the known iterative algorithms of the NIPALS type [2]. As a result, spectra of samples that are available in the library, are represented as:

$$A = S \times F + E$$

where A - n×p matrix of spectral data; n - number of spectra in the training set; S - n×f matrix of PCA-weights; p - number of points in spectrum; F - f×p matrix of eigenvectors; f - number of eigenvectors; E - n×p matrix of residual spectra.

6. For each eigenvector *i* one can calculate the average values of weights:

$$M_i = \frac{1}{N} \sum_{j=1,N} S_{ji}$$

and standard deviation:

$$\sigma_i^2 = \frac{1}{N-1} \sum_{j=1,N} (S_{ji} - M_i)^2$$

over all spectra which are included in the training set. Thus, the borders of the group for each of the orthogonal components are determined.

7. The spectrum of an unknown sample is decomposed into the obtained basis of common variations (eigenvectors) and weights S_i^* are calculated:

$$S^* = A^* \times F',$$

where S* and A* - vectors of weights and of spectral data for an unknown sample, F' - transposed matrix of eigenvectors.

8. Distance between an unknown sample and given training group is calculated:

$$D^2 = \frac{1}{f} \sum_{i=1,f} \left(\frac{S_i^* - M_i}{\sigma_i} \right)^2,$$

where S_i^* - value of the ith component of the weight vector of the unknown sample.

9. Decision taking. According to the general theory (see, for example, [3]), in the case of D>3 the unknown sample can be classified as not belonging to the given group ($P < 0.01$). In the case of D≤3 the unknown sample can be classified as the member of the given group.

4.3. MODIFICATIONS OF THE METHOD

The PCA method can be used not only to look for "similarity" (i.e. common variations) between the unknown spectrum and spectra from training set, but also to determine the degree of distinction between spectra. For this purpose we used the so-called method

PCA/MDR, which takes into account the remaining part of spectra (E - n×p matrix of residual spectra), which is not described by the decomposition of an unknown spectrum into eigenvectors of the training set. The tests showed that simultaneous use of both PCA/MD and PCA/MDR gives the best results.

However, in case of the PCA/MDR method, the presence of some unknown (i.e. not included in the training library) substances in the inspected sample can significantly increase the distance between the unknown sample and training group of substances. Therefore, to increase the reliability of the method in presence of unknown substances we have developed a method of "competing hypotheses". The essence of this method is definition of relative distances between various groups of hypotheses. For example, in case of search for explosives, it is possible to define three groups of competing hypotheses: group of explosives, group of similar in structure but non-explosive substances, and group of the background. Group of explosives can contain TNT, RDX, plastic explosives etc., the group of competing hypotheses can contain any nitrogen-containing organic substances: protein, tree root, alcohol etc; besides it, the hypothesis of background (or several typical backgrounds) is also considered. As a result of measurement and processing of the spectrum of an unknown sample, one calculates distances between the given sample and the nearest representative from each group of hypotheses. Then one can easily determine the relative probability that the unknown spectrum belongs to each of the groups. For example, the probability that the given spectrum belongs to a group of backgrounds - 94 %, to a group of explosives (TNT) - 0.5 %, and to a group of competing hypotheses (tree root) - 5.5%. In brackets we have specified substances from each group, distance from which to the unknown spectrum is minimal.

In order to increase the differences between similar substances (e.g. between TNT and a tree root) and thus to improve the reliability of the method, it is possible to introduce weighting of spectra. The weighting function for the given pair of hypotheses would look like:

$$P_i = \frac{\left|f_i^{root} - f_i^{tnt}\right|}{\sqrt{F_i^{bgd} + \left|f_i^{root} - f_i^{tnt}\right|}},$$

where i enumerates channels in spectra, f_I^{root} and f_I^{tnt} are response functions to root and TNT respectively, and F_i^{bgd} is the spectrum of the background. It is clear that such a weighting function has maxima is those parts of spectra, which are the most informative for separation of the given pair of hypotheses.

5. Comparison of LSM and PCA/MDR

5.1. APPLICATION OF LSF AND PCA/MD TO SIMULATED SPECTRA

We have carried out comparison between how LSF and PCA/MD methods can deal with the simulated spectra of TNT and tree root. Response of the experimental setup to three samples in the air – 500g of TNT and 500g tree root, 500g of melamine – were simulated by a Monte-Carlo code, assuming measurement time of 5 minute (100 spectra for each sample). The code used the real background from 5-minute measurements in the air.

Approximating response functions to C, O, N, Fe, Cl, and Si for LSF were calculated by the same code. Only n_O/n_C and n_N/n_C ratios were taken into consideration. Both methods were given pre-defined information (clusters) only about two hypotheses: TNT and tree root. All other substances were unknown. Figure 2 shows the results of recognition of simulated spectra of TNT and tree root.

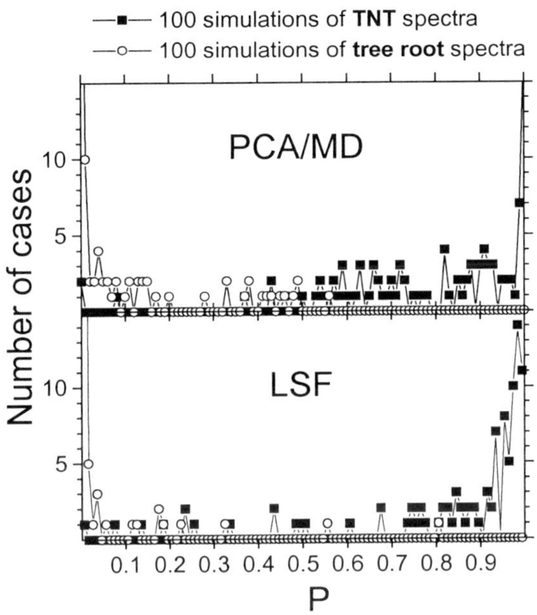

Figure 2. Distribution of probabilities that the unknown spectrum belongs to TNT.

The first impression from the Figure 2 is that LSF generally produces more definite results that PCA/MD (fewer cases of probabilities close to 50%). However, if one counts the number of mistakes (when the probability of the alternative hypothesis is higher than the true one), then results of PCA/MD start to look more reliable than of LSF. Table 1 lists the results of recognition for the two methods: percentage of misidentified TNT spectra, and percentage of false alarms (tree root spectra misidentified as TNT).

Table 1. Results of recognition of simulated spectra of 500 g TNT and 500 g tree root in open air. Simulated measurement time was 5 minutes.

	Percentage of misidentified TNT	Percentage of false alarms
LSF	12	2
PCA/MD	7	1

Of course, the above example is only an illustration, and does not include many possible refinements to both methods. It was intended for demonstration of relative capabilities of both LSF and PCA/MD to recognize simulated spectra in ideal conditions: identical response functions to TNT and root were used in simulation of spectra and construction of the corresponding hypotheses, no systematical errors that are always present in real-life experiments, and most important – no surrounding material (unknown types of sand, soil, etc), presence of which makes recognition procedure a much more challenging task.

In these ideal conditions both LSF and PCA/MD give comparable results, which can be easily improved by simulating, say, 10-minute long measurements instead of 5-minute long ones.

5.2. APPLICATION OF LSF AND PCA/MD TO EXPERIMENTAL SPECTRA FROM TNT PHANTOM IN SAND AND SOIL.

Real measurements in presence of large amount of surrounding material (soil) differ from the ideal case in several respects. Firstly, gamma-background increases due to scattering of neutrons and γ-rays from the source on the soil. The shape of this additional portion of the background may differ from the "normal" background in air and depends on the type of the soil, i.e. the effect is non-linear. Secondly, hidden object "replaces" some portion of the soil, and the resulting concentrations of elements reflect not the net values, but the difference between the object and the equivalent volume of the removed soil. Since the soil may be non-homogeneous, one cannot entirely rely on the knowledge of the background from measurements in other locations.

Experiments were conducted in the laboratory using the testing area with two types of soils: dry sand (humidity ~0.4% by weight), and black fertilized soil (humidity around 20% by weight). The exact elemental composition of the black soil was not known. The following samples were used in the measurements: 500g TNT imitator, 500g tree root, 500g steel block, 500g stone.

Measurements were carried out by the following scenario. At the first stage, several 10-minute-long measurements of the background for the given media were made. At the second stage, several dozens of 10-minute measurements for each sample were made. The resulting spectra were independently analyzed by LSF and PCA/MD procedures. For each measurement the decision was made: is this sample a TNT or not.

Tables 2 and 3 show the results for two methods of analysis. Numbers indicate the percentage of cases when the given sample was identified as TNT. For the "TNT" column it means just the rate of successful identifications of TNT, while for other columns it means how often the given sample would produce a false alarm. Typical error in both tables is about 5%.

Table 2. Results of recognition of 100 experimental spectra of 500g TNT, 500g tree root, 500g steel, 500g stone, and the background in different surrounding media by the LSF method. Measurement time was 10 minutes in all cases.

	TNT	Root	Fe	Stone	background
Air	95	0	0	0	0
Sand	67	10	0	0	3
Soil	35	6	0	0	39

Table 3. Same as Table 2, but for the PCA/MD recognition method.

	TNT	Root	Fe	Stone	background
Air	94	0	0	0	0
Sand	94	8	0	0	1
Soil	72	8	0	0	0

While for measurements in open air both methods produce identical results (as for the simulated data), PCA/MD handles the cases of sand and soil much better than LSF. While presence of the unknown component in the background (due to scattering, etc.) distorts the results of LSF, PCA/MD manages to certain extent to ignore this component. In case of the homogeneous dry sand, PSA/MD results are the same as for open air. Even when the soil contains considerable and un-controllable amount of oxygen (water), carbon (decayed organic material) and nitrogen (fertilizer), PCA/MD still can recognize the TNT in the majority of cases, while LSF can not distinguish TNT from the background and produces random results.

In order to increase the reliability of recognition for the existing experimental setup, one must measure for a longer time. Most of the misidentified TNT and false alarms come from the wet organic material (tree root in our case). More elaborate training algorithms are required to improve the recognition results in this case.

6. Conclusions

The LSF method has one important advantage over PCA/MD: it is much more intuitive. However, while it works well enough in ideal conditions, it is unclear how to handle cases when the background from the surrounding soil distorts the experimental data in an uncontrollable way.

PCA/MD algorithm is much less sensitive to presence of unknown substances, including nonlinear background effects. In our case it appears to be much more reliable.

There is also a possibility that other recognition algorithms may be even more suitable for our purpose than PCA/MD. Further research is needed to develop a reliable methods for identification of hidden explosives.

References

Vapnik V.N., (1995). The Nature of Statistical Learning Theory. Springer-Verlar, New York, Berlin, Heidelberg.

Cooley W. and Lohnes P., (1971), Multivariate Data Analysis, Wiley, NY

Gnanadesikan R., (1977), Methods for Statistical Data Analysis of Multivariate Observations, Wiley, NY

Sharaf M., Illman D., Kowalski B., (1986), Chemometrics, Wiley, NY

DETECTION OF EXPLOSIVE VAPOURS IN AMBIENT AIR BY ION NONLINEAR DRIFT SPECTROMETRY METHOD

IGOR. A. BURYAKOV

The Design & Technological Institute of Instrument Engineering for Geophysics and Ecology (IDE), the Siberian Branch of RAS, 3/6 Pr. Ak. Koptyuga, 630090, Novosibirsk, Russia.

1. Introduction

Gas analyzers are widely used to detect the presence of concealed explosives and explosive devices. The analyser produces an alarm signal on detecting explosive vapors. Fig. 1 shows the use of the gas analyzer provided with a vortex sampler for detection of an explosive device hidden within an untight object. A whirled air stream creates rarefaction close to the surface and ensures an intense air flow containing explosive vapors to the gas analyzer.

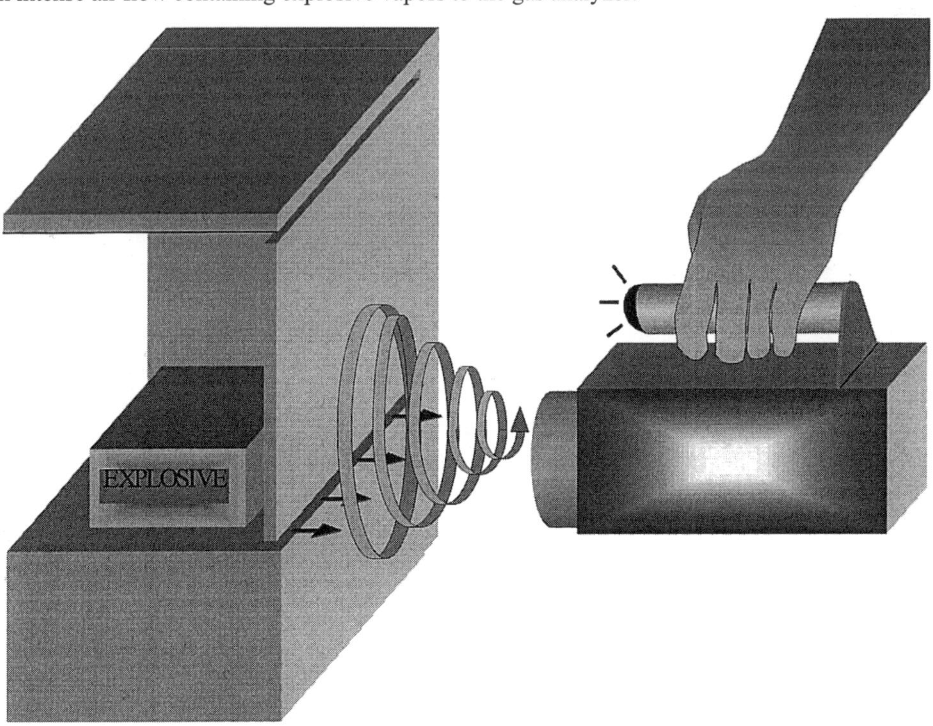

Fig. 1. The use of the gas analyzer provided with a vortex sampler for detection of an explosive device hidden within an untight object.

H. Schubert and A. Kuznetsov (eds.), Detection of Explosives and Landmines, 69–75.

High demands placed on device sensitivity and selectivity are dictated by a rather low vapor pressure of the compounds, their high adsorption power and interfering components present in air in large quantities. This is most pronounced upon solving expeditious problems on detecting trace compound concentrations in real time. The most sensitive among the well-known methods designed to detect organic traces in the atmosphere are the ones using atmospheric pressure ionization. For example, on detecting 2,4,6-trinitrotoluene the detection limit of ion mobility spectrometer is ~1 ppt [1] and of ion nonlinear drift-spectrometer (abbreviation INDS) is ~0,4 ppt [2].

INDS operation consists in sampling, sample ionization, ion separation in a carrier gas stream in a strong electric field and registration of separated ions [3-6]. A mixture of ions of different types is separated in INDS by the electric field strength dependence of the mobility coefficient. Ion drift velocity V, caused by an action of electric field is [7]:

$$V = K(E)\,E = K_0\,(1+\alpha(E))\,E, \qquad (1)$$

where E is electric field strength, K_0 is the mobility coefficient in a weak field ($E/P \leq 2$ V/(cm·torr), P-pressure), $\alpha(E)$ is a normalized function which describing the electric field dependence of the mobility.

Under the action of periodic alternating asymmetric waveform field $E_d(t)$ that meets the conditions [3]:

$$\int_0^T E_d(t)\,dt \equiv \langle E_d(t)\rangle = 0, \qquad \langle E_d^{2b+1}(t)\rangle \neq 0, \qquad (2)$$

(t is time, b is an integer, ≥1) i-type ions executing oscillatory motions with period T drift along force lines with averaged velocity V_i. This drift is compensated by constant electric field E_c. Ions for which $V_i + V_c = 0$ ($V_c = KE_c$) are transported to a detection region in a gas stream normal to the field lines, the remainder of the ions leave the region and recombine. With $V_i + V_c = 0$ compensation electric field E_{ci} is a characteristic of ion type:

$$E_{ci} = \langle \alpha E_d(t)\rangle / (1 + \langle\alpha\rangle + \langle(\partial\alpha/\partial E)\,E_d(t)\rangle) \qquad (3)$$

With E_c changing and $E_d(t)$ given spectrum of a mixture of ions of all types is recorded.

In the course of the present work a possibility of selective detection of vapours of 2,4-dinitrotoluene (DNT), 2,4,6-trinitrotoluene (TNT), pentaerythritol tetranitrate (PETN) in air with INDS has been examined and dependencies of analytical characteristics on INDS parameters have been determined. The calculated detection limit, linearity, speed of response and sensitivity of INDS on detecting vapours of the above-mentioned compounds have been determined.

2. Experimental

2.1. APPARATUS

Block-diagram of INDS is given on Fig. 2. INDS comprises an ionization chamber, an ion separation chamber, an ion detection system and a voltage generator. The ionization chamber is a heated cylindrical cavity (T_i =120°C) 1 cm in diameter and 1 cm in length. β-source ^{63}Ni with an activity of 10^{-2} Cu is placed at the inside of the chamber. Under the action of a constant electric field ions are transported between the ionization and separation chambers. The separation chamber is formed by two electrodes which are coaxial cylinders 7 cm in length, 1 and 0.68 cm in diameter (with distance between the electrodes d= 0.16 cm). Ions with purified air with flow rate Q_t=50 cm^3/s are transported through the separation chamber. An ion detection system includes an ion collector and an electrometer. To set up field in the separation chamber equal to:

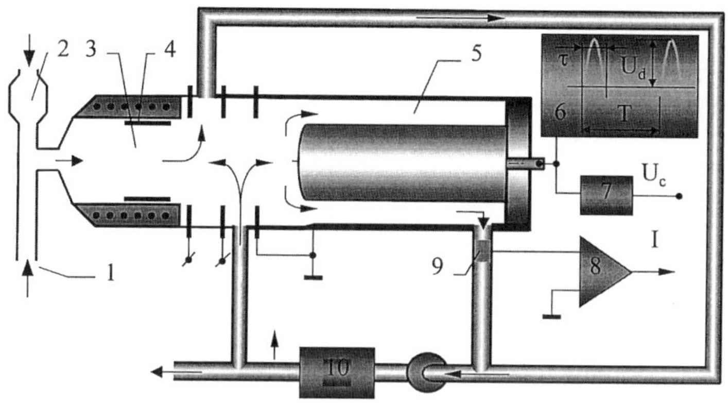

Fig. 2. Block-diagram of INDS : 1 – laboratory air, 2 – explosive, 3 – ionization chamber, 4 – β–source ^{63}Ni, 5 – separation chamber, 6 – dispersion voltage generator, 7 – compensation voltage source, 8 – electrometer , 9 – ion collector, 10 – gas filter.

$$E_d(t) = U_d \times f(t)/d, \qquad (4)$$

a waveform voltage generator was used. The generator was connected to the electrodes of the separation chamber and had the following parameters:

–voltage form - is a function normalized per unit which describes field waveform :

$f(t) = (\sin[\pi\cdot(t-mT)/\tau]-2\tau/\pi T)/(1-2\tau/\pi T)$, with $mT \le t \le (mT+\tau)$;

$f(t) = -(2\tau/\pi T)/(1-2\tau/\pi T)$, with $(mT+\tau) \le t \le (m+1)T$,

where $m \ge 0$ – is an integer. High-voltage pulse duration τ = 1.9 μs; period T= 5.9μs; dispersion voltage amplitude U_d = 1÷4 кV.

To set up compensation field E_c in the separation chamber the electrode of a small diameter was connected to voltage source U_c.

Volumetric input flow rate Q_{inp} was 5 cm^3/s. Air flow saturated with explosive vapors Q_{ex} was mixed with the input flow. To obtain Q_{ex} , purified air was passed through a quartz tube (0,3 cm in diameter, 20 cm in length) the inner surface of which was covered with the test compound. By varying Q_{ex} from 0,02 to 1,6 cm^3/s we changed the concentration of explosive vapors.

2.2. REAGENTS

During the tests doubly recrystallized explosives were used : 2,4-DNT ($C_7H_6N_2O_4$); 2,4,6-TNT ($C_7H_5N_3O_6$); PETN ($C_5H_8N_4O_{12}$).

3. Results and discussion

3.1. SELECTIVE DETECTION OF EXPLOSIVE VAPOURS IN AIR WITH INDS

On Fig. 3 drift spectrum of air containing vapours of 2,4-DNT (a), 2,4,6-TNT (b), PETN (c) under negative mode, taken with a value of dispersion voltage U_d = –3.5 kV are given. The reaction of proton dissipation yields DNT and TNT ions, and the basic reaction products are (M–

$H)^-$ ions, where M is a molecule of an explosive, H is a hydrogen atom. This is explained by the fact that DNT and TNT are gas-phase acids, and electrons are pulled to substituting groups NO_2 of a benzene ring, which causes hydrogen of a methyl group to acquire acidic properties [8]. PETN ions are produced by the reactions of electron transfer and electrophilic capture [9] with the formation of M^- or MP^-, P is a part of a reactant ion, (the main ion-reactants in air are $(H_2O)_nO^-_2$, where $n \geq 0$ is an integer).

An essential characteristic of a spectral device is resolution which is indicative of the achievable degree of compound separation. The INDS resolution can be determined by the following expression:

$$R = U_c / W_{50}, \qquad (5)$$

where $U_c = E_c(\alpha) \times d$ is a compensation voltage which classifies ion type, W_{50} is width at half-height of the ion peak on a drift spectrum. Dimensions of a separation chamber, parameters of the separation and compensation voltages, carrier-gas flow rate, ion diffusion at the electrodes effect W_{50}.

Fig. 3 shows the experimental data of resolution R for the ion peaks of DNT,TNT, PETN equal to 15, 11, 3, respectively. A low value of R for PETN ions is due to a weak α (E) dependence.

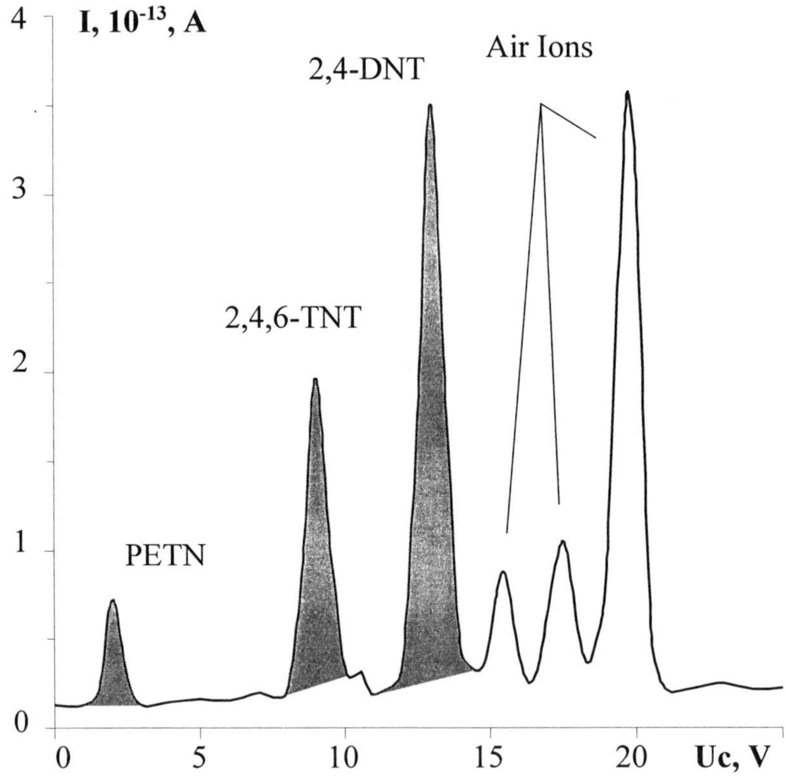

Fig. 3. Drift-spectra of air containing vapour of: a) DNT, b) TNT, c) PETN obtained under negative mode. $U_d = -3.5$ kV.

To evaluate the device possibilities to separate a binary compound mixture it is convenient to use a separation criterion employed in chromatography [10]: $Cr=(U_{c1}-U_{c2})/(\sigma_1+\sigma_2)$, where σ is a standard peak deviation ($\sigma = W_{50} / 2{,}35$ on condition that the peak follows Gaussian distribution).

The conditions of a separate detection of two peaks:

$$\begin{array}{ll} \text{no separation} & - \ Cr = 1, \\ \text{sufficient separation} & - \ Cr = 2, \\ \text{complete separation} & - \ Cr = 3. \end{array} \qquad (6)$$

Criterion $Cr > 3$, $(U_{ci}-U_{cj}) \geq 4$ V, $(\sigma_i+\sigma_j) < 0.73$ V is for any pair of explosives. Reference to Fig. 3 shows that peaks of explosives are completely separated from one another and from background. Provided that only one or several U_c values corresponding to explosives are applied, selective detection of only components concerned is possible.

3.2. ANALYTICAL CHARACTERISTICS OF INDS

The concentration dependence of the detector signal is expressed by Eq. [11]:

$$I = D \times C^k, \qquad (7)$$

where D – is a constant, k refers to sensitivity coefficient, C – is the compound concentration. Experimental dependencies of the ion current amplitude on the vapor concentration of DNT, TNT and PETN are given on Fig.4 at the logarithmic scale. The concentration dependence of the ion current amplitude I(C) as exponential expression (7), calculated detection limit (L) and the values of saturated explosive vapour concentration at room temperature and atmospheric pressure are

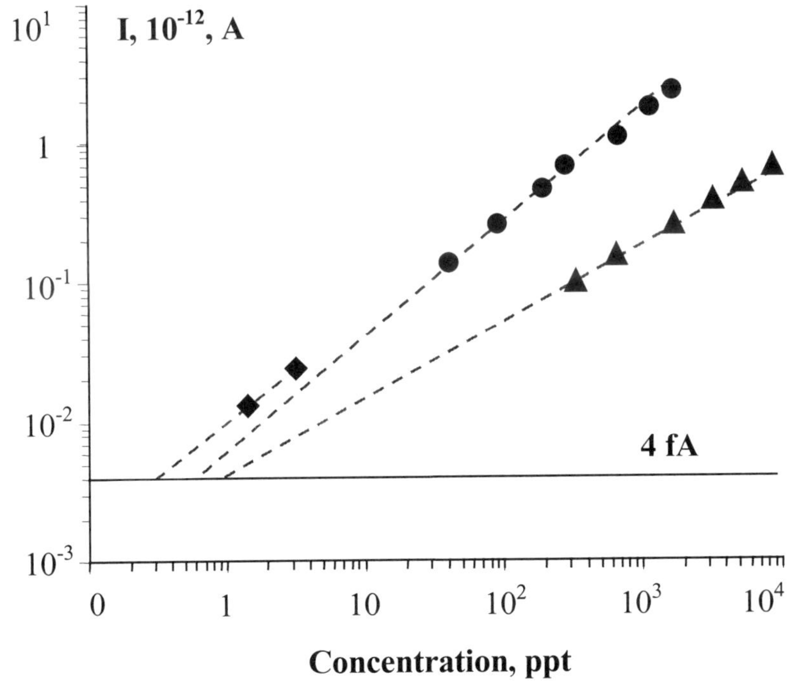

Fig. 4. Experimental dependencies of ion current amplitude on DNT (π), TNT (✯), PETN (◆) concentration.

presented in Table 1. The detection limits L of DNT, TNT were calculated from experimental dependencies (7) as a value of concentration at which ion current amplitude (I) was equal to double the width of zero line (intersection point of two lines: the line that passes through experimental points and "4 fA" line). The width of the zero line was 2 fA. The detection limit of PETN was determined directly from the experiments.

Table1. Analytical characteristics of INDS.

Compound	**I(fA) =D×C^k(ppt)**	**L, (ppt)**	**C$_{нас}$, ppb (v/v), [12]**
2,4-DNT	$I = 3 \times C^{0.6}$	2	55.7
2,4,6-TNT	$I = 12 \times C^{0.7}$	0.4	9.4
PETN	$I = 9 \times C^{0.85}$	0.3	0.018

For the detector with a linear dynamic range sensitivity coefficient k in expression (7) should meet the condition: $0.98 \leq k \leq 1.02$ [11]. Sensitivity coefficients k for DNT, TNT, PETN were: 0.6, 0.7, 0.85, respectively, with an overall measurement error of ± 0.10. This suggests that INDS response on detecting tested compounds is not linear. Hence, for these compounds using INDS one can through relation C_{max}/L (C_{max} is the maximum injected compound concentration) determine the width of the dynamic range alone. With the given k values the range width was over 10^3on detecting DNT and TNT.

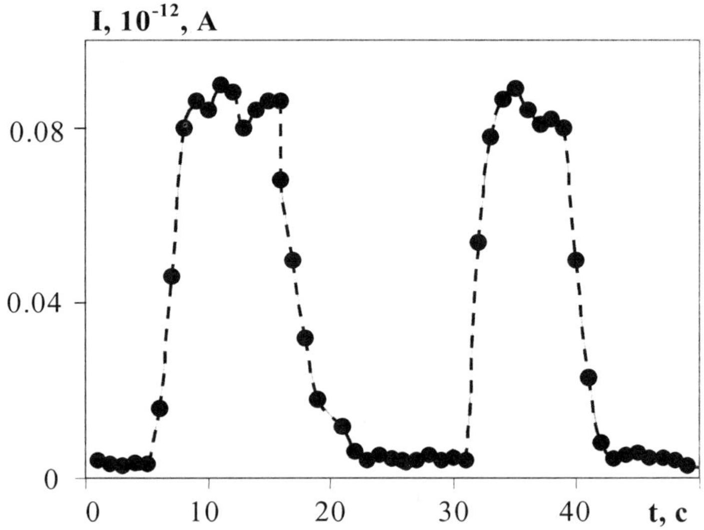

Fig. 5. INDS response to varying concentration of TNT (300 ppt) in air.

3.3. SPEED OF RESPONSE (t_R)

Speed of response (t_R) is determined as the time required for the device to obtain 0,9 of complete signal variation amplitude. In INDS t_R is determined by time t_t required to transport ions to the collector and electrometer speed t_e. In theory t_t value is given by the volumes of the input lines, ionization and separation chambers, Q_{ex}, Q_t , and for the device used $t_t \approx 0.6$ s. $t_e \approx 0.2$ s. During the experiments on determining the INDS time response to TNT with U_c=8.7 V t_R value of about 2 s was obtained (Fig.5). Aftereffect time t_a=4÷5 s. Such high t_R value appears to be due to residual sorption of TNT by the input lines, which is responsible for an additional delay upon the sample delivery to the ionization chamber.

4. Conclusion

The results of testing have shown a possibility for selective detection of samples containing 2,4–dinitrotoluene, 2,4,6–trinitrotoluene, pentaerythritol tetranitrate with INDS. Calculated detection limit, linearity, sensitivity and speed of response of INDS on detecting explosives vapour are determined. The results of experiments point to high sensitivity, selectivity and speed of INDS response. These properties of INDS render it as the most promising device for expeditious detection of explosives.

5. Reference

1. *Cohen M.J., Wernlund R.M., Stimac R.M.* // Nucl. Mater. Manage (Proc.Issue). 1984. V.13. P. 220.
2. *Buryakov I.A., Kolomiets Y.N., Louppou V.B.* // Int. J. for Ion Mobil. Spect., 4 (2001), 1, pp. 13-15.
3. *Горшков М.П.* // А.С. №966583 СССР. 1982. Б.и. №38
4. *Буряков И.А., Крылов Е.В., Макась А.Л., Назаров Э.Г., Первухин В.В., Расулев У.Х.* // Жур. Ан. Химии. 1993. Т. 48. В. 1. С. 156.
5. *Buryakov I.A., Krylov E.V., Nazarov T.G., Rasulev U.Kh.* //Int. J. of Mass Spectrom. and Ion Processes. 1993. T. 128. P. 143.
6. *B. Carnahan, S. Day, V. Kouznetsov, A. Tarassov.//* Proc. of 4th Int. Workshop on IMS, pp. 789-800 (1995).
7. *Mason E.A., McDaniel E.W.* Transport Properties of Ions in Gas. New York: Wiley, 1988. 560 p.
8. *Spangler G.E., Carrico J.P., Campbell D.N.* // J. of Testing and Evalution, JTEVA. 1985. V.13. № 3. P. 234.
9. *Danylewych-May L.L.* // Proceedings of the First Int. Symp. on Explosive Detection Technology, Atlantic City. 1992. P. 672.
10. E. Leibnitz, H.G. Struppe. *Handbuch der Gaschromatographie*. Akademische Verlagsgesellschaft Geest & Porting K.-G. Leipzig, 1984, p.47.
11. *Scott R.P.W.* Liquid Chromatography Detectors. Amsterdam: Elsevier, 1986. P.28.
12. V.C. Dionne, D.P. Roundbehler, E.K. Achter, R.J. Hobbs, and D.H. Fine. Vapor pressure of explosives. *J. of Energetic Materials.* **4:** 447 (1986).

Trace detection of components emanating from hidden explosives

Dr. Peter Kolla, Bundeskriminalamt (Federal Police Agency), Germany

Keywords:
vapour detection, explosives, trace detection, EVD, sniffer technologies, canines

Abstract:
Different strategies for explosives detection are discussed. A special focus is set to sniffer technologies and the range of substances emanating form explosives that have to be detected. Technical sniffers are compared to canines with respect to sensitivities and type of detected substances.

Discussion:
The search for hidden explosives as it is done for security reasons is based on two different strategies. The quickest way especially to check suspicious objects by technical means is the use of penetrating radiation and a typical response is detected. Mostly applied techniques are based on x-ray technology.
The other strategy relies on the emission of characteristic components from the explosives and/ or the contamination of the environment of the explosives by characteristic components. These components are detected technically using trace analytical equipment or besides pure technical equipment by especially trained dogs.
The substances features should allow selective detection i.e. the substances are typical ingredients of explosives. In addition to the uniqueness, the substances should have functional groups, that enable selective detection in low amounts and the physical properties must allow the collection/ sampling of these substances to transfer them to the detection systems.
Looking generally at substances in explosives or explosives devices the following types can be found wherein usually only the energetic substances are really typical:

- Energetic substances (nitro-organic explosives, nitrates, peroxides etc.)
- Fuels (mineral oil, vegetable oil, aluminium, sugar, sulphur etc.)
- Residues from manufacturing processes (solvents, by-products, starting material etc.)
- Plasticizers, stabilisers and phlegmatising additives (polymers, oils, waxes, phthalates, esters etc.)
- Wrapping material (foils, paper, plastic containers etc.)

The mostly used instrumental methods for trace detection are high-speed gas chromatography coupled to specific detectors and ion mobility spectroscopy. Both methods are very sensitive and can be operated highly selective to explosives detection. The range of components from explosives is reasonably the explosive substances itself. This restriction is necessary in order to reduce false alarms and to increase the reliability of a positive explosive detection. Depending on the individual system the explosive substances from the above list to be detected are AN, EGDN, NG, TNT, PETN and RDX. Explosives that not contain any of these substances are not detectable by usual equipment.
The most often discussed problem in the application of instrumental trace detectors is the kind of sampling. The sampling procedure strongly influences the success of the analytical method. Looking at the explosive substances the biggest difference in the physical properties from the sampling point of view is the vapour pressure.

H. Schubert and A. Kuznetsov (eds.), Detection of Explosives and Landmines, 77–81.

	Equilibrium vapour pressure at room temperature [Torr]	Concentration in air [ppt (v/v)]
EGDN Nitroglycol	4.8×10^{-2}	63 000 000
NG Nitroglycerin	2.6×10^{-4}	340 000
2.6-DNT Dinitrotoluene	2.8×10^{-4}	370 000
2.4-DNT Dinitrotoluene	5.0×10^{-5}	65 000
TNT Trinitrotoluene	4.0×10^{-6}	5 300
PETN Pentaerythritol tetranitrate	2.2×10^{-8}	28

Whereas EGDN and NG have rather high vapour pressures the volatility of PETN and RDX is very low. Therefore the sampling procedure can not only rely on vapour detection. The sampling method must also effectively collect particles and droplets.
The classical vapour analysis of EVD's (explosive vapour detectors) as well as canines work with the gas composition above the explosives. Looking at the gas phase composition most explosives show obvious differences to the bulk composition.

Explosives - Composition of gas phase and bulk

Example 1: Propellant (double base NC):

Bulk composition:
main components : nitro-cellulose, nitroglycerin
by-components : stabilizers, phlegmatizers
(z.B. diphenylamine, centralite, phthalate, dinitrotoluene)

Gas phase composition:
main components : acetone, toluene, nitroglycerin
by-components : limonene, phthalate

Example 2: Composition C4:

Bulk composition :
main components : hexogen
by-components : polyisobutylen, cyclohexanone

Gas phase composition :
main components : cyclohexanone
by-components : toluene, 2-ethyl-1-hexanol, cyclopentane

Example 3: Trinitrotoluene (depending on manufacture):

Bulk composition :
main components : 2,4,6-trinitrotoluene
by-components : 2,4-dinitrotoluene

Gas phase composition :
main components : 2,4-dinitrotoluene
by-components : 3,5- u. 1,4-dinitrotoluene,
1,3- u. 1,2-dinitrobenzene,
2,4,6-trinitrotoluene

The gas phase compositions as given above are usually measured in the vapour above the explosive in an equilibrated container. The vapour of a freshly prepared explosive device depends strongly on the package, temperature and venting conditions. Even under favourable conditions near an equilibrium state, it takes some time for the explosive constituents to penetrate thin package and form a steady vapour cloud.

Experimental results of gas phase concentration of explosives:

Test setting: Sampling of the vapour outside a dynamite and a military TNT

- each packed in LD-PE bags (thickness 0.2 mm)
- vapour sampling with EGIS-collector above a not covered box for 20 s
- sampling after periods of minutes to hours
- evaluation of detector response (peak area)

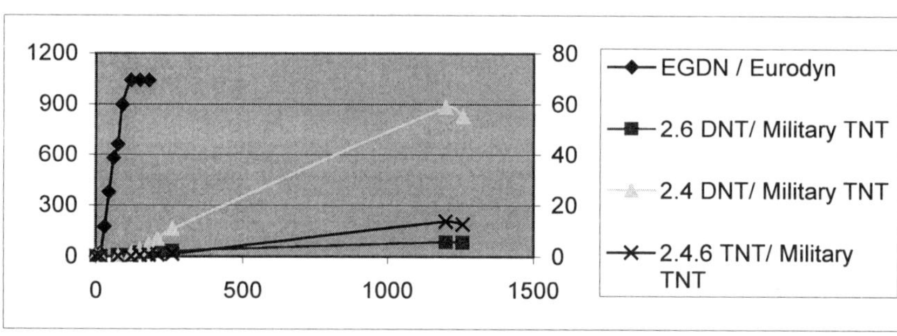

Legend:
y-axes: left- area units of EGDN
right- area units of DNT‘s and TNT
x-axis: period in minutes

The results of testing the concentration of the explosive substances in the vapour after different periods of time show the fast penetrating behaviour of EGDN from dynamites. Looking at the TNT sample, obviously the 2,4-DNT emanates mcuh faster and in a higer concentration form the bulk as the TNT itself. Even the 2,6-DNT, whose amount in the military TNT-block is very low (< 0.25 %) can be detected in a distinct concentration.

The use of trained dogs seems at first sight to be similar to the application of technical trace detectors with the only difference in the kind of detection principle but a closer look at the detection work done by dogs shows many differences.

	Trace Analyzer	Canine
Qualitative substance detection	highly selective nitro-compounds	broad range nitro-compounds, by-products, additives and combinations
Sensitivity, limit of detection	0.1 – 1 ppb	0.1 – 100 ppb
Dependence on concentration	fixed limit of detection, mostly linear response to concentration, overloading alarm	individual, environmental change of limit of detection, detection gaps, no "overloading" alarm"

Dogs do not only detect a close range of explosive substances and have also a different sampling method compared to technical trace detectors. They are well suited for fast searches of not too large areas.
From technical point of view the most obvious problem of the canine is the lack of output of an analytical result showing qualitative and quantitative information.

Conclusion:
The use of substance analytical methods for explosives detection depends essentially on substances emanating from the explosives. These substances must be unique and analysable. The analytical systems applied as detector whether technical equipment or trained canines need substances that enable the clear recognition of an explosive as origin.
In the real world environment the amount of substances and range of concentration of different substances makes the clear recognition a complex problem.

References:

P. Kolla, "Anwendung analytischer Verfahren zur Detektion versteckter Sprengstoffe – das Aufspüren von Sprengvorrichtungen", Angewandte Chemie, 109, S. 828-839, 1997

J.M. Johnston et al., "Canine Olfactory Capability for Detecting NG Smokeless Powder, 5th Int. Symp. Anal. Det. Expl., Washington, Pap. 43, 1995

P. Kolla, Riechen Hunde Sprengstoff, Polizei, Verkehr, Technik, 2000, 4/00, pp. 110-114

P. Kolla, Stability of Explosives Traces on Different Supports Respecting the Detectability by EVD, Proceedings of the SPIE Symposium, Chemistry and Biology Based Technologies for Contraband Detection, Boston, 1996, Vol 2937, pp. 236-244

Landmine Detection with an Electronic Nose Mounted on an Airship

J. Goschnick, M. Harms
Forschungszentrum Karlsruhe, Institut für Instrumentelle Analytik
Postfach 3640, D-76021 Karlsruhe

Abstract

An Electronic Nose (EN) might be a promising alternative to sniffer dogs for detecting landmines on the basis of their characteristic gas ensemble. The latter does not only consist of the gas released by the explosive itself but to a larger extent of the gases produced by the explosive´s so-called by-products and mine packaging. The Karlsruhe Micronose KAMINA is a novel type of EN developed at the Karlsruhe Research Center. It has primarily been designed for use in consumer applications. The gas sensitivity of the thumbnail-sized microarray chip is based on a single semi-conducting metal oxide thin film. Its low price, weight and small size as well as its low power consumption combined with the appropriate analytical power provide an excellent basis for mine detection. The KAMINA mounted on a 9 m long radio-controlled airship allows the location and classification of objects with characteristic gas release such as landmines. First test results obtained with tin dioxide microarray chips detecting contaminated soil model samples showed that when using a 2 m long sniffing tube a maximum speed of 3 km/h was appropriate for a safe detection of gas emitting objects. When simply moving the KAMINA system a few centimeters above the ground (i.e. without using the sniffing tube) a maximum speed of even 6 km/h was found to be feasible. The spatial resolution of the micro-pump sampling was about 10 cm. The newly developed microarray with a nanogranular layer of tin dioxide showed an extremely high gas sensitivity. Laboratory tests with model gas exposures revealed detection limits of less than 1 ppb. Thus, microarray chips of this type provide an excellent basis for the delicate task of gas analytical landmine detection. The cheap and easy maneuverable radio-controlled airship seems to be an ideal platform for moving the EN over the terrain to be investigated.

Introduction

The so-called odor bouquet of a landmine is nothing else than a characteristic gas ensemble which is detected by sniffer dogs when they are employed for mine detection outside in the field. In view of some draw-backs of sniffer dogs (such as limited daily working time, limited age span dogs could be used for landmine sniffing, easy deception of dogs noses) artificial olfaction with an Electronic Nose (EN) might be an alternative. On the other hand the successful employment of dogs demonstrates that the gas release of mines is a valuable marking even when the landmines are buried in the ground.

Landmine screening means a rapid and nevertheless thorough examination of large areas that are likely to contain landmines. Firstly the mine-free areas should be identified and secondly, the mines located as quickly as possible. Of course the time and costs involved are also important issues that have to be considered. Therefore, a low-cost gas analytical technique should be chosen that is capable of a fast detection and of focused sampling on the ground. Spatially resolved screening is achieved by rastering. At this point a small radio-controlled airship comes into play as an inexpensive platform for moving the gas analytical device over the ground. The advantages of an electro-motor driven airship are manifold: its exhaust does not interfere with the ambient gas composition. It is easy to maneuver, can operate near ground without touch down and if necessary it can stop to allow a more detailed analysis than during travelling. Vibrations can be held very low avoiding disturbances of the measurements. Last but not least, the costs are low compared to other hovering platforms such as e.g. helicopters.

Sophisticated conventional instruments may be able to perform an on-line analysis of low concentrations, but they are too spacious and heavy for a mobile operation in the field – not so

H. Schubert and A. Kuznetsov (eds.), Detection of Explosives and Landmines, 83–91.

an EN. Further important to mention is that the odor bouquet only contains a small amount of the explosive itself. TNT usually used in antipersonnel landmines and the explosive´s even more volatile by-products [1] only contribute to a lesser extent to the gas signature released by the mine. It are mainly the painting, printing dyes, machine and anticorrosive oils as well as plastic parts that determine the vapor ensemble in the vicinity of the mine. Therefore, sensing the vapor of the explosive itself might be extremely difficult, especially when considering that the explosive is encapsulated while the painting is outside. Consequently, vapor sensors focusing on the gas release of the explosive encounter an extremely difficult job to do.

An EN with its broad sensitivity spectrum seems to be more appropriate because all components of the mine's gas signature contribute to the EN's signal output. It works like a dog's nose, detecting the mine via an integral recognition of the gas suite released by the mine. Unfortunately, commercial EN's suffer from similar drawbacks concerning robustness, spatial requirements and weight as the conventional analytical instruments. Moreover, most of them cannot be operated on-line and the price of more than 50.000 US$ is a considerable hindrance to their use.

A novel type of EN has been developed at the Institute of Instrumental Analysis of the Karlsruhe Research Center which has been designed to meet the requirements of consumer products: low price, small size and weight as well as low power consumption combined with appropriate analytical power [2]. The Karlsruhe Micronose (Karlsruher Mikronase) KAMINA is based on a unique microarray of gas sensors mounted in a sensor module the size of a beverage can but nevertheless contains all necessary electronics. The microarray chip utilizes a new type of an array consisting of thin film metal oxide (MOx) gas sensors. The electrical conductivity of these gas sensors responds extremely sensitively to the ambient air composition. The sensitivity spectrum is very broad since only chemically very passive gases cannot be detected such as rare gases or nitrogen. Because the sensitivity spectra of the sensors in the array are brought about by a so-called gradient technique (see below) each sensor reacts differently to gas exposure. This different behavior results in a characteristic signal pattern which allows recognition and discrimination of single gases or gaseous mixtures. Thanks to the simplicity of the sensing principle and the structure of the microarray the device is inexpensive, light weighted and robust.

The KAMINA was mounted on a 9 m long radio-controlled airship in order to evaluate the potential of this technique for locating and identifying objects with characteristic gas release, such as gas leaks in a pipeline, odorous materials at a dump site or sources of gaseous pollutants. Antipersonnel mines maybe another type of object to be located and recognized. First experiments were made to investigate the basic performance of the moving electronic nose to odor plume tracing.

Instrumental

The thumbnail-sized gradient microarray chip (see Fig. 1) allows low-cost fabrication as all sensor elements are produced in one step by partitioning a single sputtered MOx film with parallel electrodes to measure the electrical conductivity of the sensor elements formed by each of the metal oxide segments with its adjacent two electrode strips [3]. The standard microarray chip used in the reported investigations comprises 38 sensor segments of platinum doped tin dioxide and 4 rear meander heaters to enable an operation at 200 – 400 °C. Elevated temperature is needed to allow fast and reversible signal response of the metal oxide sensors. A gradient technique is used to differentiate the sensor segments in terms of their selectivity for gas response. A temperature variation across the microarray maintained by inhomogeneous heating with the 4 heaters and an additional thickness gradient of a gas permeable SiO_2 coating on top of the metal oxide are employed to individualize the gas response of the sensor segments. A beverage can-sized device (see Fig. 2) has been developed containing the complete EN system including the microarray chip, a micro fan for sampling and the µp-controlled electronics, providing heat regulation and continuous sensor data acquisition with 1 Hz. The instrument has a high discrimination power for gases or odors and allows fast response times of a few seconds. For most organic as well as inorganic gases detection limits in the range of 100 ppb were obtained.

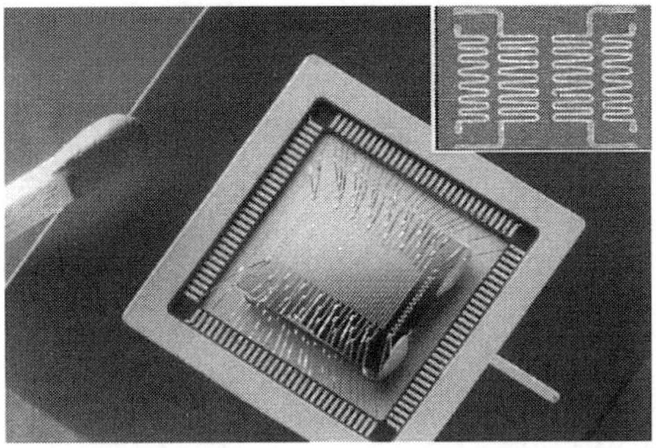

Fig. 1: Gas sensor chip with a gradient microarray mounted in its housing. The chip is based on a segmented metal oxide film of which the electrical conductivity is dependent on the ambient gas composition. Two thermoresistor strips beside the metal oxide film serve as temperature control. Electrical contacts are provided by thermosonic gold wire bonds. The rear side of the chip (upper right) carries 4 meander-shaped heating elements made of platinum to allow inhomogeneous heating of the chip.

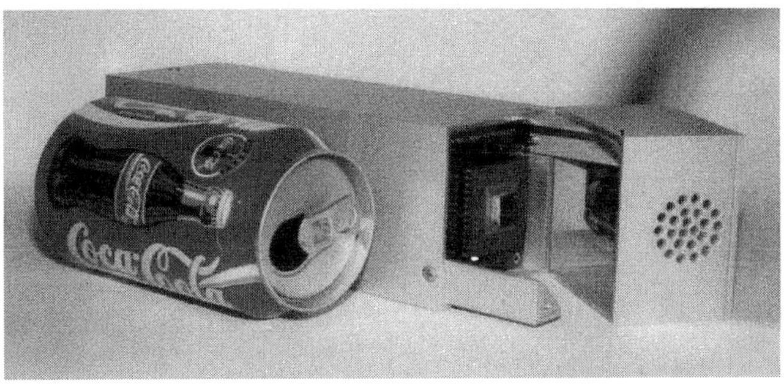

Fig. 2: KAMINA containing the complete electronic nose system including the microarray chip, a micro fan for sampling and the µp-controlled electronics.

This electronic nose was installed on a radio-controlled airship which has a GPS positioning system and a video camera on-board. A board computer and a data transmission unit record the data and send them down to the ground station where the gas analytical data as well as the position data are available in real time. The zeppelin system (length: 9 m, width 2.6 m in the middle) has just performed its first flights successfully. A sniffing tube of aluminium (2 m length, 14 mm internal diameter) goes down to the ground which enables focused sampling (see Fig. 3). A raster flight course of the airship over the area to be investigated and continuous data acquisition by the KAMINA allow us to record odor maps of the area and find objects of specific gas signature such as landmines.

Fig. 3: Airship LENA (**L**uftschiff mit **e**lektronischer **Na**se) equipped with a sniffing tube to enable focussed sampling (left), a video camera and the KAMINA (right)

Experimental

First tests of the gas analytical equipment were performed with the KAMINA and the sniffing tube for sampling only, i.e. without using the airship. For evaluating the feasibility of screening analyses spatial resolution an area of 3 x 4 m contaminated with two different gas sources was rastered by following a path with distances of 1 m between the path ways (s. Fig. 4). The gas sources were glass dishes of 9 cm in diameter filled with soil one soaked with ethyl acetate while the other contained soil contaminated with ethanol. The dishes were immediately opened before starting the test. Other tests were carried out to determine the maximum speed at which a sufficient sensor signal can be obtained by the KAMINA being moved over a gas source. For the latter a sample of 100 µl ethanol on a tissue was used. The KAMINA with the sniffing tube was moved with different speeds above the sample (s. Fig. 5). During all tests the funnel-shaped opening of the tube was placed ca. 10 cm above the vapor sources and the sampling flow was adjusted by a micropump attached at the KAMINA head to be 0.75 l/min. Moreover, in another set of tests measurements at different distances between vapor sources and tube funnel were performed. The same tests were repeated without the sniffing tube simulating an assembly where the KAMINA device is carried far below the airship moving it in short distance above ground. So two different types of sampling techniques could be compared: sampling with a sniffing tube while the KAMINA was fixed to the airship bottom and direct sampling with the KAMINA simply lowered to the ground.

Furthermore for enhanced gas sensitivity, a new type of gas sensor microarray with a nanogranular layer was tested with n-butanol exposures. This test was carried out with an automated pulsed gas exposure system. In order to take the dynamic behavior of the gas sensor system into account, pulsed exposure series were performed in which the chip after clean air exposure was abruptly exposed to air containing a definite concentration of a model gas. The latter was applied with six different concentrations (0.1, 0.2, 0.5, 1, 5, and 10 ppm) at constant humid air of 50 % r.h. After a certain time the chip was flushed with clean humid air and subsequently exposed stepwise to the next test pulse containing another concentration of the model gas.

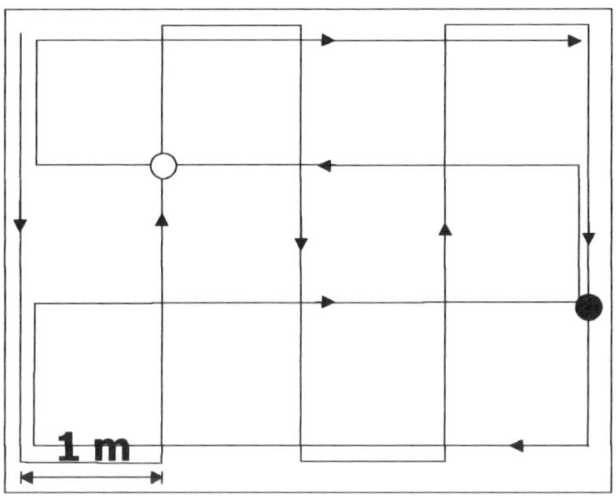

Fig. 4: Raster course to test detection and distinction between two gas sources (ethanol, grey, and ethyl acetate, black). The start is depicted in the above left-hand corner.

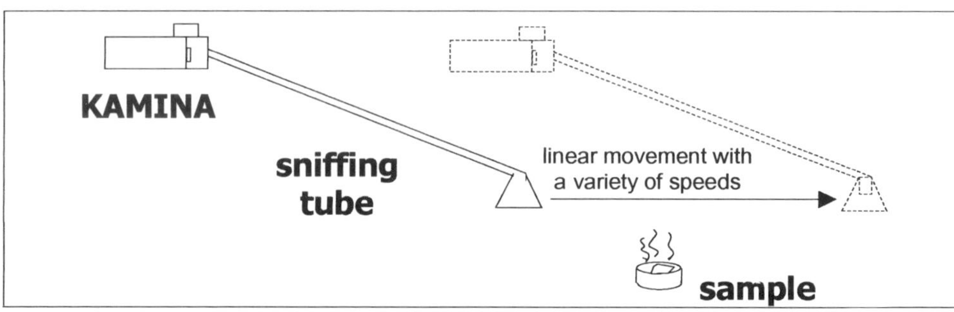

Fig. 5: Set-up of KAMINA with sniffing tube tested for its response speed

Results

The resistances of the 38 sensor segments of the gradient microarray showed a typical stepwise decrease versus time when the gas release of the model sources started to reach the microarray (see Fig. 6). Moreover, due to the selectivity differences in the gas response of the array elements resulting from the two gradients across the array the resistance patterns change when the air composition above the microarray changes, e. g. because of ethanol evaporation. The gas characteristic patterns of the 38 resistances are depicted in polar plots, where each beam represents one sensor segment. The signals shown are the relative conductivity changes with respect to clean air and normalized to the median of all sensor segments. So the pattern of clean air forms a circle, whereas contaminated air results in characteristic patterns which allow gas recognition.

The signal pattern produced at ethanol exposure is shown in Fig. 6 during tests with the sniffing tube. There was a delay of 20 s till the gases reached the microarray surface causing the sensor resistances to decrease. Without the tube the response time was only 1 s. According to the tube travel time of about 20 s the gas molecules can spend some time with adsorption at the tube surface which cause an increased delay and a tailing of the sensor signals beyond the end of exposure. At a lateral displacement of 1 m between tube end and gas source, the sensor signal decreases by a factor of 2, at 1 m vertical displacement by a factor of 7 and at lateral and vertical displacement simultaneously by a factor of 4. Obviously, the plume of the gas source does not only move upwards but rather seems to move diagonally. The maximum speed to detect clear signals above the significance level (triple noise amplitude) was found to be 3 km/h using the sniffing tube. Without the tube moving the KAMINA at low height along the ground the maximum speed was found to be 6 km/h. In both cases the samples were 100 µl ethanol soaked tissues put in cylindrical petri dishes of 9 cm in diameter.

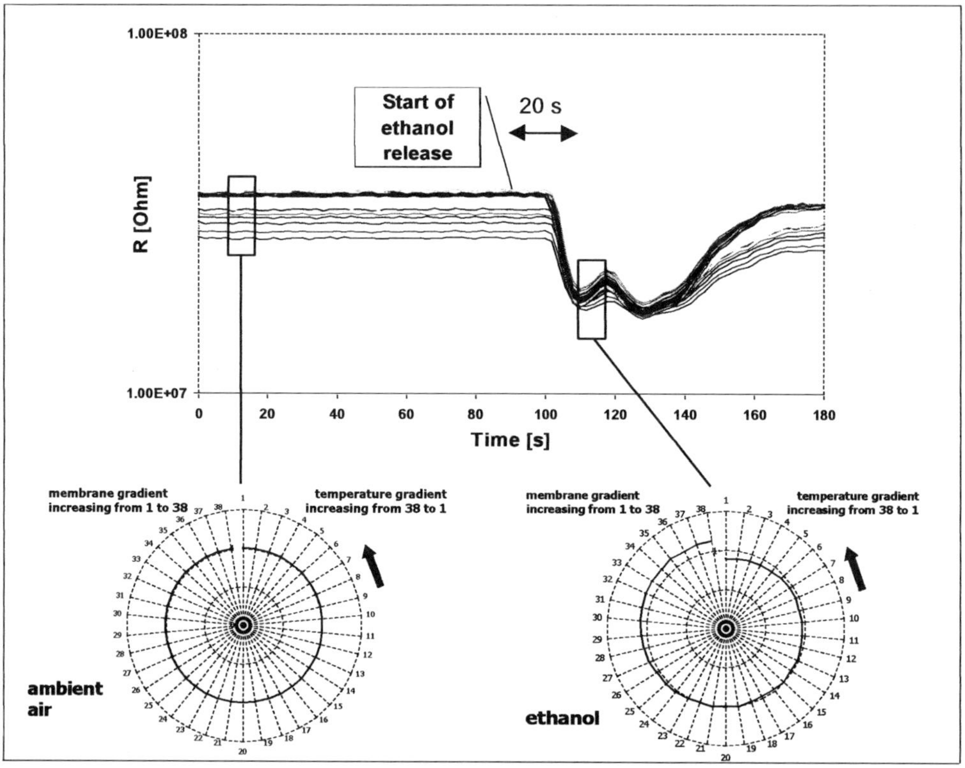

Fig. 6: Resistance versus time for all 38 sensor elements of a test with an ethanol vapor source at a distance of 10 cm between source and tube funnel (upper plot) The polar plots show the normalized signal patterns obtained before and after release of ethanol vapor. The sniffing tube causes a delay time of 20 s.

The analytical results of a raster course above two vapor sources (ethanol and ethyl acetate) is depicted in Fig. 7. The funnel of the sniffer tube was moved with a speed of 0.35 km/h at a height of 10 cm. A Principal Component Analysis (PCA)[1] shows two separated data clusters because of the gas distinction power of the microarray: data clusters of ethanol and ethyl acetate are clearly separated from the area of ambient air. After making up for the travel time delay in the tube the distribution of the two gases was transferred on the scheme of the raster course. The grey and black lines indicate recognition of ethanol or ethyl acetate by the KAMINA. Both gas sources were detected without any problems and could be assigned to the right kind of vapors because of the characteristic signal pattern. However, whereas ethanol was only detected behind the gas sources directly, ethyl acetate was also analyzed in some parts of the rastered area far away from the source. There are two probable reasons. The gases passing through the sniffer tube temporarily adsorb at the inner walls of the tube. Because strength and extent of the adsorption are dependent on the kind of gas, different delays in the travel time through the tube were found for different gases. So the adsorption of ethyl acetate was obviously stronger than that of ethanol. Moreover, the spatial distribution of the gases changed during the course of the experiment. Initially the sample dishes were opened and the vapors were concentrated around the position of the sample dishes. After some minutes, especially at the second run of the raster course (s. Fig. 4), the gases spread over the rastered area in left direction from the original sample position because of little draught. Tests without the sniffing tube confirm this interpretation. Without using the tube the gases were also analyzed at some distance from the sample position in the later stage of the test, but the time delay correction was not necessary any more. The sampling technique with as well as without the tube, but with the KAMINA lowered to the ground seems to be sufficient to transport gases for an analysis to the KAMINA. Because of the gas dependent delay time due to adsorption direct sampling without the sniffing tube has to be preferred.

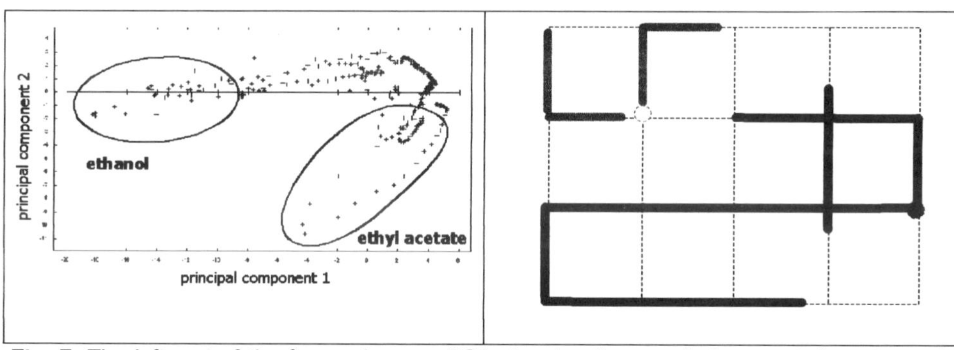

Fig. 7: The left part of the figure shows a PCA signal pattern analysis recorded continuously during rastering. The right part of the figure shows the raster course along which the KAMINA including its sniffer tube was moved. The height of the sniffer funnel was 10 cm above the ground, while the speed of the movement was 0.35 km/h. The area covered contained two different vapor sources marked at the flight path diagram (grey: ethanol, black: ethyl acetate). See Fig. 4 to compare with the direction of the raster course.

However, the standard microarray metal oxide layer might not be sensitive enough to detect such delicate gas signatures as released by landmines. In order to be well equipped for the

[1] The Principal Component Analysis is a frequently used mathematical method to investigate signal patterns of sensor arrays. The PCA is an optimum projection of the data from the original signal pattern space build up by as many dimensions as sensor elements are in the array (here 38 sensor segments) to the two dimensional diagram shown [4]. Each location in such a PCA plot represents a certain signal pattern: Similar patterns are sitting next to each other. The more different the signal patterns are the greater the distance between them. Due to the scatter in the measurements "same" signal patterns form a more or less narrow data cluster while repeated measurements of significantly different signal patterns yield two clearly separated data clusters.

detection of objects with only faint gas release a novel, highly gas sensitive metal oxide film has been developed. With the help of a plasma-based nanoparticle deposition process a thin film of nanogranular tin dioxide is placed on the chip substrate and subsequently processed in the usual way dividing the microarrays into 38 segments.

These microarrays with a nanogranular gas detection layer showed an high sensitivity to the test gas n-butanol. Detection of gas concentrations in the lower ppb range is clearly feasible and the extrapolation of the measured calibration curves to the triple noise level of the electrical measurement (about 10% relative conductivity change) even suggests detection limits below 1 ppb (see Fig. 8). The calibration curve of a conventionally sputtered gradient microarray at 300 °C is also shown for comparison. The extrapolated curve intersects the triple noise level magnitudes higher, namely at approximately 100 pbb. Even with a temperature of 150 °C the microarray with the nanogranular layer is still more sensitive than the ones with sputtered tin dioxide layers. A detection limit below 10 ppb can be achieved. Consequently, the use of lower operation temperatures allows to save a considerable amount of energy without any loss of gas sensitivity.

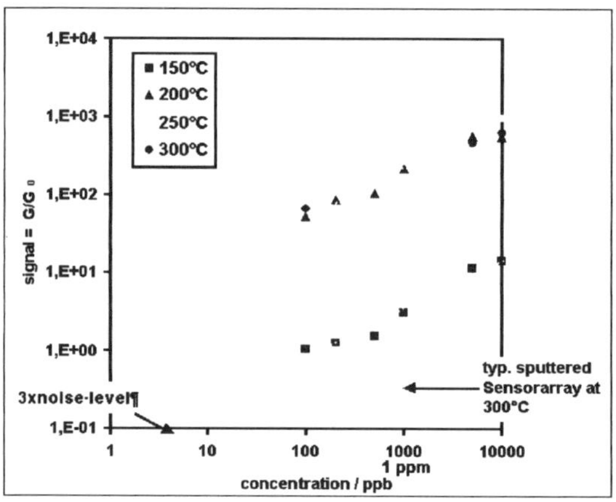

Fig. 8: Calibration curves of a microarray chip with a nanogranular layer of tin dioxide at different operation temperatures. The high sensitivity in comparison to typical sputtered microarrays with sputtered tin dioxide layers can be clearly seen.

Conclusions

The Karlsruhe Micronose KAMINA equipped with a microarray of a metal oxide nanogranular layer provides an excellent basis for the delicate task of gas analytical landmine detection. KAMINA microarrays of this kind offer a very high sensitivity to model gases in laboratory tests with detection limits lower than 1 ppb. Continuous sampling is feasible, as no pre-concentration is necessary at such low detection limits. Moreover, the microarray has proven to have a high discrimination power for different gases. So a recognition of the mine gas signature seems feasible even in environments of other gas emittents. The low weight of 250 g and its high robustness allow a mobile use in the field.

The airship LENA impressively shows the advantages of such a platform for gas analytical object search missions such as landmine detection. No touch down to the analyzed terrain is necessary. LENA is easy to maneuver and the height and motion speed can be reduced to zero if necessary. Moreover, the application of electro motors prevents exhaust gases from disturbing the analysis and propeller induced gas mixing can be avoided.

Odor maps of areas under suspect seem to be feasible with a spatial resolution of some 10 cm. The first experiments suggest a maximum cruising speed in the search mode of 1 – 3 km/h using the 2 m long sniffing tube and even 6 km/h with the KAMINA lowered to the ground. However, there is still a long way to go until a reliable mobile gas analysis instrument for mine detection is ready for application. Further experiments especially under conditions of high similarity with the practical application will be necessary to gain more insight into how the instrumental details have to be designed in order to achieve analytical results of high quality and great reliability. Hence, numerous experiments in well-controlled simulated scenarios and in the field with real mines need to be performed in order to adapt the instrument to practical conditions.

References

[1] V. George, T. F. Jenkins, D. C. D. C. Leggett, J. H. Cragin, J. Phelan, J. Oxley, J. Pennington, Progress on Determining the Vapor Signature of a Buried Landmine, Proc. SPIE, Detection and Remediation Technologies for Mines and Minelike Targets IV, Vol. 3710, Part 2, (1999) 258

[2] P. Althainz, A. Dahlke, M. Frietsch-Klarhof, J. Goschnick, H. J. Ache, Reception Tuning of Gas-Sensors Microsystems by Selective Coatings, *Sensors and Actuators B*, 24-25 (1995) 366-369

[3] P. Althainz, J. Goschnick, S. Ehrmann, H. J. Ache, Multisensor Microsystem for Contaminants in Air, *Sensors and Actuators B*, 33, (1996) 72-76

[4] K. R. Beebe, B. Kowalski, An Introduction to Multivariate Calibration and Analysis", Analytical Chemistry, 59, No. 17 (1987) 1007A-1017A

DETECTION OF SEMTEX PLASTIC EXPLOSIVES

By: Petr Mostak, Miroslav Stancl
Research Institute of Industrial Chemistry
Aliachem a.s. Division Synthesia
53217 Pardubice-Semtin, Czech Republic

Abstract

Methods of detection of plastic explosives were evaluated and the effectiveness of the detection considered. Basic tools of this detection are electronic detectors, X-ray systems, dogs and colour reactions. The background why plastic explosives are rather good detectable is explained .
The direct and indirect contamination of surfaces by plastic explosives was studied and characterised. The marking of plastic explosives for detection is discussed. 2,3 dinitro-2,3 dimethylbutane is the most popular marking agent, which is broadly used. The detectability of this compound is not very good by some detectors.

1. Introduction

Detection of plastic explosives is an important task in the security measures against criminal misusing of this type of explosive. The danger of plastic explosives is usually connected with high explosive strength, easy forming and relatively low vapour tension of explosive components of these explosives. The plastic explosives were used in many criminal bombings, the attacks on civil aeroplanes at which the big number of passengers was killed and these criminal events shocked the international community.
The criminal misuse of plastic explosive SEMTEX H stimulated legislative activities in the Czech Republic. Czech Ministry of Industry and Trade strictly forbade the export of all plastic explosives in 1989. The specialists of the Czech Republic took part in the work of the Ad Hoc Group of experts which prepared as the special body of ICAO the technical part of the Convention on the Marking of Plastic Explosives for the Purpose of Detection. This Convention was agreed by the International Conference in Montreal on the March 1991. The Government of the Czech Republic ratified this Convention in 1991 as one of the first countries.
Scientists of the Research Institute of Industrial Chemistry developed very quickly a technology of marking of plastic and sheet explosives in accordance with requirement of the Technical Annex of Montreal Convention. All plastic and sheet explosives produced in the Czech Republic have been marked for detection from June 1991 and only marked plastic explosives can be exported under strict licence treatment performed by the Ministry of Industry and Trade.

The process of industrial production of 2,3 dinitro-2,3 dimethylbutane was developed in our Institute and this marking agent has also been transported to some other countries.
The measures mentioned above removed any possibility to sell the unmarked plastic and sheet explosives in the Czech Republic or export these products after 1991.

H. Schubert and A. Kuznetsov (eds.), Detection of Explosives and Landmines, 93–102.

Taking into account the fact that the Convention entered into power as late as in 1998 and the presence of unknown stocks of unmarked plastic explosives in various countries the question of detection of unmarked plastic explosives remains a matter of high interest. This presentation is trying to consider some aspects of this problem.

2. Composition of Plastic Explosives

The main explosive components of plastic explosives are RDX and PETN. Sometimes a mixture of RDX and PETN is used; both explosives are used in SEMTEX H and also in some types of military plastic explosives.
The second important component of plastic explosives is a binder which makes the product consistent having plastic or elastic properties. The binder is usually composed from suitable polymeric substance and plasticizer. Polyisobutylene, polystyrene, polyacronitrile, or polyethylene are often used as polymers; the usual plasticizers are dioctylphtalate, dibutylphtalate or dioctylsebacate.

3. Detection of plastic explosives

The detection technologies used for detection of plastic explosives are very similar, very often identical with procedures and equipment used for detection of other explosives. The reason is that detection of plastic explosives is based on detection of the main explosive components it means RDX, PETN or both. The other components of plastic explosives are not suitable for detection purposes.
There are many methods which can be used for detection of plastic explosives, this paper is focused on methods based on trace analysis which are most frequently used in practical applications. Therefore, the systems detecting explosives by specific bulk properties of explosives are not the subject of this presentation. We have in mind such systems as enhanced X-ray, neutron based systems, quadruple resonance, computing tomography or MM wawe imaging.
The trace detection is based on the following possibilities:

- vapour detection
- particles detection
- combination of vapour and particles detection

3.1 Vapour detection

This method of detection is dependent on the vapour tension of explosive components of plastic explosives.
In fact the vapour tensions of RDX and PETN are rather low especially at RDX in comparison with such explosives as TNT, NG or EGDN. This is the reason why the detection of plastic explosives by vapour detection is more difficult than the detection of other explosives as dynamites, ammonites, TNT and further explosives containing NG or TNT. The low vapour pressure has been also the reason why plastic explosives including SEMTEX were considered as non-detectable explosives and this aspect was broadly stressed by mass media. This argumentation is not correct, the performance of detection techniques has improved in last years dramatically and best detectors are ready to detect plastic explosives by vapour detection. Pre-concentrators are often used to raise the detection probability.
The most effective electronic detectors are using IMS and GC systems, some of them the combination of both.

3.2 Particles detection

Detection of particles is an alternative to vapour detection of explosives. It was found that contamination of surfaces and persons by particles of explosives takes place at any manipulation with explosives. The most often way of contamination by particles are fingerprints done by hands of people working with explosives, further important way is spreading of the particles in surroundings by forming and manipulation with explosives or dispersing particles from contaminated shoes, suits and transport means.
The particle detection opened the excellent possibility to detect effectively the explosives having a low vapour pressure. In fact plastic explosives are very good detected by particle detection because the mass of traces fixed by fingerprints on surfaces are fully sufficient for detection. This is caused by sticky property of plastic explosive; the consequence is the massive contamination of hands which are source of good detectable fingerprints.
The fingerprints from plastic explosives and substantial contamination of surfaces by indirect contamination done at forming of plastic explosives are studied in the part 4 of this paper.
Taking into account the sensibility limits of up-to-date detectors which are usually in the range of 20-200 pg it is clear that 1 particle is sufficient for successful detection. It is the reason why particle detectors are very popular in spite of the longer time, which you need for detection in comparison with vapour detection.

3.3 Combination of vapour and particles detection

The combination of vapour and particles detection could be the effective system for detection of plastic explosives. Some of the detectors are now able to work in two modes, one is detecting vapours, second particles. Unfortunately both modes cannot work at this time simultaneously because the particles detection needs the collection of particles by vacuum exhauster or swabs and this operation is time taking procedure which requires thermal desorbtion of explosive traces from swabs or preconcentrators.

3.4 Detection by colour reactions

Detection of explosives by colour reactions is easy and quite effective procedure how detect the particles of explosives. This method uses the chemical reactions with explosive by which the compound having specific colour is formed. This colour is typical for individual explosive or groups of explosives. The information obtained is not so exact as in case of detectors based on GS or IMS procedures used in vapour and particles detection and the sensitivity of colour detection is lower in comparison with IMS or GC.
On the other hand the method is very simple and cheap and even at the lower sensitivity one particle of explosive found is usually sufficient for positive detection.
The detection of plastic explosives is influenced to some extent by the presence of binder on the surface of explosive particles, which defend the access of reaction solutions to particles and therefore, the development of relevant colour takes longer time.
The detection set DETEX II developed in our Institute is used by Czech police, Army and is exported to other countries. The experience obtained proved the usefulness of this method. The procedure of detection using colour reactions can be seen from the Table 1. Similar detection systems were developed and have been used in other countries.

Table 1 Colour reactions of explosive compounds

Groups	Compound	Detex I	Detex II	Solution No	Colour reaction
Nitroaromates	TNT	*	*	1	Dark violet
	DNT	*	*	1	Green to blue
	Tetryl	*	*	1	Orange
	Tetryl	*	*	1+2+3	Pink
	Picric acid	*	*	1	Yellow
	Ammonite	*	*	1	Dark violet
Nitramines	RDX	*	*	1+2+3	Pink
	HMX	*	*	1+2+3	Pink
	Semtex	*	*	1+2+3	Pink
	C4	*	*	1+2+3	Pink
Organic nitroesters	Dynamit	*	*	1+2+3	Pink
	NG	*	*	1+2+3	Pink
	EGDN	*	*	1+2+3	Pink
	DINA	*	*	1+2+3	Pink
	PETN	*	*	1+2+3	Pink
Inorganic nitrates	NH_4NO_3		*	2+3+4	Pink
Inorganic chlorates	$KClO_3$		*	5	Blue

3.5 Detection by dogs

The good-trained dogs are an effective means of explosive detection. The dog detection ability and sensitivity can be compared with electronic detectors. Dogs are able to detect the hidden plastic explosives with high probability of positive result. The disadvantage of dog is a relatively short time period in which the dog is able to perform a detection work. This time is usually 20-30 min., than the animal needs to have a rest. Anyway, in a short time action the dog is excellent detector, his nose has the sensitivity fully comparable with the best electronic detector. It was proved in searching tests for hidden explosives on the board of aircraft that the probability of explosive detection is on the similar level for dogs and electronic detectors and the combination of both gives some synergy and best result.

Taking into account the information mentioned above, it is clear that the view often published in mass media that plastic explosives are not detectable is not correct and this type of explosive can be detected by all detection method specified above. It is true that the detection of plastic explosives by vapours is not so effective, the low vapour pressure of PETN and RDX in comparison with other explosives causes this fact.

4. Contamination level in trace contamination of various surfaces

Two modes of work were used in the contamination tests:
- direct contamination by fingerprints
- indirect contamination by mechanical forming of plastic explosive

Explosives used: SEMTEX 1A (PETN based plastic explosive)
SEMTEX H (RDX based plastic explosive
Samples of materials: aluminium
Polyethylene
Leatherette
Sample size: direct contamination - thin plates 5x25cm
indirect contamination - thin plates 15x15cm
Analytical and detection methods: DETEX II
IONSCAN 400
HPLC

4.1 Direct contamination

Fingerprints on the testing plates were prepared by the thumb, which was contaminated by waddling motion of the thumb firmly, pressed in the explosives. Ten fingerprints were prepared by imprinting of the contaminated thumb onto the surface of testing plate.
Sampling for detection was performed by wiping of the fingerprint using at the detection by IONSCAN special textile swab, at DETEX II filter paper on which the colour development is than applied, the a swab performed by filter paper was used for extraction of explosive traces for HPLC. Results of detection of fingerprints prepared from SEMTEX H and SEMTEX 1A are presented in the table 2.

Table 2 Detection of fingerprints of SEMTEX H and SEMTEX 1A

Explosive	Carrier	Sample	Fingerprint No.					
			Blank	1	2	3	5	10
Semtex 1A PETN	Aluminium	IONSCAN	Pass	10,900	9,500	9,200	9,000	6,700
		DETEX	-	++	+	++	+	-
	Polyethylene	IONSCAN	Pass	11,500	11,200	9,300	11,200	10,000
		DETEX	-	+	+	+	+	-
	Leatherette	IONSCAN	Pass	11,500	11,300	11,400	8,800	6,900
		DETEX	-	+	-	-	-	-
Semtex H RDX	Aluminium	IONSCAN	Pass	ovld.	11,200	11,500	11,700	11,500
		DETEX	-	++	++	++	+	+
	Polyethylene	IONSCAN	Pass	ovld.	ovld.	ovld.	11,100	11,300
		DETEX	-	++	++	+	+	+
	Leatherette	IONSCAN	Pass	ovld.	ovld.	ovld.	ovld.	11,600
		DETEX	-	++	++	+	+	+

Note: - the values presented at IONSCAN present the cumulated amplitude of peaks of explosive component, ovld. means overloaded
- at DETEX II detection where we obtain only relative values means:
- ++ positive detection
- + slightly positive detection
- - negative results

The results presented in the Table 2 are demonstrating that fingerprints of SEMTEX H and SEMTEX 1A can be easy detected on the various surfaces by particle detection systems. It can be also seen from the results that DETEX II can easier detect SEMTEX H than SEMTEX 1A and the influence of surface material on detection of SEMTEX 1A was found.
The important question is what mass of explosive contains a fingerprint and what part we are able to gain for analysis. These values are dependent on more factors including adhesion of plastic explosive to the skin of the thumb, adhesion to the tested surface, properties of swabbing material and procedure of taking swabs. To obtain some basic information concerning content of explosive in fingerprints prepared in our tests on plates the HPLC analysis of explosive content in swab and unwiped rest of explosive was done. The thumb with and without glove prepared the fingerprints. The SEMTEX H was used in these experiments, the results obtained are presented in the Table 3. The values presented are the mean of 3 measurements.

Table 3 RDX content in fingerprints prepared by SEMTEX H on polyethylene

Average RDX content				
Explosive	**Sample**	**Original fingerprint**	**Wipe (swab)**	**Rest (rinsing)**
Semtex H	Blank 1	0	0	0
	Glove	3,135 ng	2,334 ng	978 ng
	Blank 2	0	0	0
	without glove	2,303 ng	1,124 ng	1,155 ng

It can be seen that in the tests performed 50-70 % of explosive present in fingerprints was gained by swabbing, it is evident that this way of collecting of explosive traces is very dependant on various conditions.

4.2 Indirect contamination

The indirect contamination is defined as the contamination of persons, walls and further objects near to the place where any manipulation with explosives was performed. The spreading of particles can be influenced by mechanical working of explosive, air movement, movement of contaminated persons or other means.
The model test of indirect contamination used the following procedure:
200 g of plastic explosive was moulded (tearing and recombining) above the testing plate 15x15 cm in the distance 5 cm from the plate surface for 3 min. The contaminated surfaces were sampled by touch, wiping and rinsing and the presence of explosive was detected by IONSCAN 400, DETEX II and analyzed by HPLC. The results of these tests are presented in the Table 4.
The results obtained proved that in the process of plastic explosives forming big number of particles is dispersed in surroundings and contaminates the surfaces. The real quantity of explosive on tested plate is much higher than values of HPLC analysis presented in the Table 4 because the extract for analysis was prepared from only one touch of swab with the surface of the tested plate.
This contamination can be transported by air movement in relatively long distances and cause a general contamination of room where some operation with explosive was done. Suit and shoes can transfer such contamination to other places. When we take into account that one particle of explosive could be sufficient for successful detection, the contamination by particles mentioned above is a good chance for forensic work.

Table 4 Indirect contamination of surfaces by forming of plastic explosives

			Type of collecting			
Explosive	Carrier	Sample	Blank	Touch	Wipe	Rins.(ng)
SEMTEX 1A PETN	Aluminium	HPLC	0	not perf.	not perf.	7,993
		IONSCAN	-	8,700	ovld.	ovld.
		DETEX	-	+	+++	not perf.
	Polyethylene	HPLC	0	not perf.	not perf.	14,069
		IONSCAN	-	ovld.	ovld.	ovld.
		DETEX	-	++	+++	not pef.
	Leatherette	HPLC	0	not perf.	not perf.	2,318
		IONSCAN	-	ovld.	ovld.	ovld.
		DETEX	-	++	+	not perf.
SEMTEX H RDX	Aluminium	HPLC	0	not perf.	not perf.	2,794
		IONSCAN	-	9,500	ovld.	ovld.
		DETEX	-	+++	+++	not perf.
	Polyethylene	HPLC	0	not perf.	not perf.	3,799
		IONSCAN	-	ovld.	ovld.	ovld.
		DETEX	-	+++	+++	not perf.
	Leatherette	HPLC	0	not perf.	not perf.	1,374
		IONSCAN	-	ovld.	ovld.	ovld.
		DETEX	-	++	+++	not perf.

Note - ovld. means overloaded
- not perf. means not performed

4.3 Discussion

It can bee assumed that contamination done by fingerprints or dispersion of particles of plastic explosives at manipulation with explosive is rather massive giving a good chance of detection of presence of this type of explosive.

The results presented in Table 2 has given clear evidence that the traces of SEMTEX H and SEMTEX 1A can be easy detected by IONSCAN 400 on various surfaces, even the 10th fingerprint was detected in all cases. Also the detection by colour reaction was successful especially at fingerprints of SEMTEX H. The 10th fingerprint was detected in most experiments. This result was surprising taking into account fact that the sensitivity of colour reactions is approx. 100x lower than is achieved at IONSCAN 400. Such result can be explained by the ability of colour reaction detection to react with the whole mass of explosive collected on the swab. At detection using IONSCAN 400 there are substantial losses of explosive in thermal desorption and in the drifting tube. Therefore the real difference in sensitivity of both systems is not so high.

The mass of explosive in the SEMTEX H fingerprints prepared on PE was, as can be seen from the Table 3, on the level 2000 –3000 ng what is in relation with results of P. Neudorfl (1) who found the mass of the first fingerprint prepared by SEMTEX H on stainless steel 4595 ng and of the10th fingerprint 135 ng. We found in experiment with SEMTEX 1A the explosive mass of the 10th fingerprint 80 ng.

On the other hand in the same work mass of fingerprints on PE was only 500 –700 ng.

In the work of M. Langford (2) much lower explosive mass in fingerprints was found on various surfaces as can be seen from following examples:

Al	4 ng RDX	18 ng PETN
Plastic	26 ng RDX	55 ng PETN
Cloth	40 ng RDX	50 ng PETN

It is clear that the results of experiments with fingerprints are influenced by several conditions, which cannot be effectively standardised. The main factors is adhesion of plastic explosive to various surfaces and physical properties of explosive binding system.

Important question is how much of explosive mass we are able to gain at collecting of explosive traces by wiping or exhausting by vacuum. L. Elias (3) studied the effectiveness of swabbing by cotton swab and by flexible nickel gauze. He found that 85-90 % of explosive contained in fingerprints can be collected.
The similar results have been achieved in the work of R.T. Chamberlain (4) who achieved by so called Dry Transfer Method the gain of 90 % of explosive mass of fingerprint.
As can be seen from the Table 3 we achieved the effectiveness 50 –70 %. Anyway, it can be considered that the effectiveness of the explosive traces collecting is quite good even when we take into account the probability that the collecting in operational condition should be less effective.

The experiments with indirect contamination performed by forming of plastic explosives proved that at manipulation with plastic explosives substantial quantity of explosive particles are dispersed in surroundings.
We can see from the Table 4 that the mass of explosive which was found was in the range of 1400 – 8000 ng. This corresponds to approx. 20 ng/cm^2.
It is known from other tests that some particles are transported by air movement to longer distances and usually contaminate other objects near the place where was manipulated with explosive. A secondary transfer of contamination by explosive particles adhered on shoes, suits or hands of persons is also well known and was studied for instance in the NATO CRG (5).
Taking into account the sensitivity of detectors of explosives which less than 1 ng of explosive and the fact that the mass of 1 small explosive particle(3-4 microns) is approx.50 ng it is clear that the presence of plastic explosives can be effectively detected by traces which are formed by direct and/or indirect contamination.

5. Marking of plastic explosives for detection

The Convention to the marking of plastic explosives was accepted in 1991 and entered in force in 1998.The Technical Annex of the Convention specified the 4 marking agents and their concentrations to be used for marking of plastic explosives.
In praxis only DMNB and p-MNT are used. EGDN and o-MNT cannot be used from hygienic and safety reason. Most popular marking agent is DMNB, some countries increased the DMNB content in plastic explosives to 1 % to secure the required presence of DMNB in explosive during the rather long shelf life.
Some problem has been evolved with detection of DMNB vapours by IMS. In fact the marking agents were tested and proposed in the Technical Annex in 1989-1990 when the best detectors used GC as detector EVD 1 or EGIS. Therefore, the marking agents fit especially to this system. The further development of detectors was oriented mainly on IMS which has

some advantages over GC. The quicker detection, lower dimensions and lower price can be mentioned for instance. Unfortunately the sensitivity to DMNB and p-MNT is rather low. The low sensitivity to DMNB explained A. H. Lawrence in his study of the thermal stability of cluster ions of dinitroalkanes. It was found that the Cl adduct of DMNB has the thermal stability only 50°C. The low thermal stability of this cluster ion is the reason of high losses in drifting tube resulting in bad sensitivity of IMS detectors. On the other hand the thermal stability of some other dinitroalkanes is much higher, the stability of 1,4-dinitrobutane is as high as150°C. This leads to the idea to look for an additional marking agent which could be better detected by IMS.
The another reason for such assessment is the relative toxicity of DMNB and p-MNT. The further unsuitable fact is that 2-nitropropane which is the realistic basic material for industrial production of DMNB is produced only by one company in the United States and therefore the presence of this product on market and reasonable price is not sure.

6. Conclusion

Plastic explosives can be effectively detected by various detection methods including detectors using GC and IMS systems, colour reactions and special trained dogs.

The detection is based mostly on particle detection, the collection of one particle is usually sufficient for positive detection.

The trace contamination having the form of fingerprints or dispersed particles evolved at forming or manipulation with SEMTEX plastic explosives produces sufficient quantity of explosive for successful detection. The mass of plastic explosive in fingerprints is dependent on various conditions, usually the first fingerprint contains several thousands ng, the tenth fingerprint approx. ten times less.

The results obtained at indirect contamination demonstrated that substantial contamination is taking place in the process of plastic explosive forming. Therefore the probability that the person manipulating with plastic explosive charge will contaminate the surroundings is high.

DMNB and p-MNT are used as marking agents for improving the vapour detection of plastic explosives. The lower sensitivity of IMS to these compounds presents the difficulty at some detection systems. The search for some more suitable marking agent for IMS is a theme for consideration.

7. Literature

[1] P. Neudorfl, M. A. Mc Cooeye, L. Elias
Testing protocol for surface testing detectors
Proceedings of the 4th International Symposium on Analysis and Detection of Explosives, September 7-10, 1992, Jerusalem, Israel

[2] M. Langford
Report in the Proceedings of the 4th Workshop on the Detection of Explosives, October 19-22 1999 Ascot, Berkshire, UK

[3] L. Elias
Swab sampling of PETN and RDX deposits on various surfaces
4th Workshop on the Detection of Explosives, October 19-22, 1999 Ascot, Berkshire, UK

[4] R.T.Chamberlain
Development of explosive standards on cloth material
4th Workshop on the Detection of Explosives, October 19-22, 1999 Ascot, Berkshire, UK

[5] F. Volk, A. Cummings, M. Horacek, P. Mostak
Investigation of Sites Contaminated by Explosives and Propellants Ingredients
Final Report to the NATO CRG, February, 1999

[6] A. H. Lawrence, P. Neudorfl, J. A. Stone
The Dependence of Cluster Ion Formation on Molecular Geometry with particular Reference to Dinitroalkanes and Chloride Ions
6th International Symposium on Analysis and Detection of Explosives, July 6-10, 1998 Prague, Czech Republic

HPM/IR Detection of Landmines: Review and Update

J.S. Seregelyi, S.M. Khanna, R. Apps, M. Boyle & F. Paquet

Defence Research Establishment Ottawa,
Ottawa, ON, Canada, K1A 0Z4

1. Abstract

A new hybrid remote-sensing method, using active high-power microwave (HPM) illumination and passive infrared (IR) detection, is being developed at the Defence Research Establishment Ottawa (DREO) for the detection of shallowly buried landmines.

A 2.45 GHz microwave source was used to illuminate a landmine buried in soil. The thermal signature of the mine at the soil surface was detected in the 8-12 μm region both in near real-time, as well as after a brief time-delay following illumination. The thermal signature at the soil surface is primarily made up of two components. One component occurs in near real-time due to the interference of the incident beam and the energy scattered by the mine. A second component is generated when the heating pattern, due to differential microwave absorption by the mine and the surrounding soil, is conducted upwards from the mine location to the soil surface. Both signatures are dependent on the complex dielectric constants of the mine and the soil. The method is forward-looking to avoid the risk of straddling the mine during detection, and it can be used on mines with little or no metal content.

In this review, a brief description of the method will be given. Laboratory and field trial results for the detection of; non-metallic mine surrogates, dummy mines without explosives, and real mines with explosives but without a fuse will be presented. Results of various numerical studies will also be presented. To the best of our knowledge, the present data are amongst the first unambiguous imaging results using HPM/IR detection of real mines.

2. Introduction

It has been estimated that there are about 110 million anti-personnel (AP) and anti-tank (AT) mines scattered on the ground surface or buried in the ground in about 64 countries[1]. These mines pose a serious threat to both unsuspecting civilian populations and military operations: including UN peace-keeping operations. In addition, the effect on the local economy is often devastating, as is the case in Afghanistan, Bosnia etc. A mined area can never be safe until it is thoroughly cleared of mines.

The recent international treaty to ban the use of antipersonnel mines by most countries of the world has provided a significant push to eliminate these weapons from the arsenal

H. Schubert and A. Kuznetsov (eds.), Detection of Explosives and Landmines, 103–124.

of mankind, and a welcome support to the cause of demining. Unfortunately, modern mines contain very little metal and are difficult to detect using conventional electromagnetic techniques. As a result, there are currently about 20 methods for mine detection at various stages of development[2]. They range from quite simple methods, such as the use of dogs, to the most sophisticated modern techniques including the use of; nuclear quadrupole resonance, thermal neutron activation, acoustic techniques, magnetic measurements using superconducting quantum interference devices (SQUIDs), and chemical detection methods.

For military as well as civilian/humanitarian applications, a successful mine detection method must be inexpensive, easy to use, and should have a fast and accurate detection rate. Further, any new detection method must have superior detection sensitivity, detection rate and a lower false alarm rate than already available through existing methods. The method should also be forward-looking to avoid the risk of straddling the mine during detection.

Amongst the various detection methods under development, passive infrared imaging[3], conventional electromagnetic methods, ground probing radar (GPR), and thermal neutron activation[4] (TNA) are perhaps the most promising techniques. Hyper-spectral imaging is also expected to yield a powerful method for mine detection. While these methods also have their respective limitations, a fusion of data from these sensors could provide a system that may be acceptable for most applications. Amongst these methods, passive IR imaging[3] is particularly attractive due to the simplicity of the technique, remote-sensing capability and relatively low cost as compared to the other methods. However, this method has its own problems. In this technique, the mine signature is strongly dependent on the diurnal variations in solar illumination, type of soil, soil moisture content, and temperature gradient in the soil. A mine signature may be almost non-existent under cloudy conditions. Active infrared methods have been proposed for mine detection. These methods typically require the use of a scanning laser system and reflections from the mines (on the ground surface) could provide information on the location of the mines. Recently, Li et al.[5] have obtained results that indicate the possibility of imaging surface and buried mines through the use of thermal step function excitation using infrared heating lamps. Limited target signature, background clutter and false alarms under various experimental conditions are the principal problems in these methods.

In Feb. 96, a new method was suggested for mine detection at the Defence Research Establishment Ottawa (DREO). The method was based on a hybrid remote-sensing technique involving active high power microwave illumination and passive detection in the infrared wavelength region. Hybrid remote sensing involving active illumination in one wavelength and detection in another wavelength is not new but, in general, the detection and illumination wavelengths regions are not too different in wavelength. It was later discovered that this method was also being investigated independently by two other research groups [6,7,8,9]. Our results and interpretation of this technique have both similarities and marked differences from their descriptions of this method.

In this paper, a brief description will be given first of the method involving active high-power microwave illumination coupled with passive infrared detection of targets buried in the ground. The difference between the HPM/IR method and enhanced surface heating coupled with IR imaging will be discussed. Laboratory and field trial results for; the detection of non-metallic mine surrogates, dummy mines without explosives, and real mines with explosives but without a fuse will be presented. Results of various numerical studies will also be presented. To the best of our knowledge, the present data are amongst the first unambiguous imaging results for detection of real mines with this method.

3. The HPM/IR Detection Method [10,11]

The complex dielectric constant of a mine and the soil surrounding it are, in general, different. As a result, when a mined area is illuminated with microwave radiation, a temperature variation at the soil surface above the mine (relative to the nearby soil surface) is observed. This temperature contrast can be detected with an infrared imaging system and may be exploited for mine detection.

At this point, it is instructive to review the passive IR imaging technique for mine detection. Simard[3] has given a clear description of the processes involved in passive IR imaging of buried mines. As described by Simard,[3] these processes lead to different soil surface temperatures, T_m and T_r, above the mine and at a nearby soil surface reference site without a mine, respectively. Assuming a one dimensional conduction process, these temperatures can be determined by solving the differential heat equation for the three zones a, b and c shown in Figure 1a.

$$\frac{\partial T}{\partial t} = K_T \frac{\partial^2 T}{\partial x^2} \tag{1}$$

Here T is the temperature, t is time, and x is depth in the soil and K_T is thermal heat diffusivity. It is assumed that the soil surface is illuminated uniformly and there is uniform heat flow Q from the atmosphere into the ground surface above the mine and the surface surrounding the mine. Simard[3] has discussed a simplified model for the above process. The soil column above the mine and an equivalent soil column in a reference region with no mine have different heat impedance Z_m and Z_r respectively, as shown in Figure 1b. A uniform heat flow Q is assumed from the atmosphere into these two adjacent regions. This would lead to different soil surface temperatures T_m and T_r above the mine and in the reference (no mine) region. $(T_m\text{-}T_r)$ gives the target thermal signature which is observed by thermal imaging. Clearly $(T_m\text{-}T_r)$ is proportional to Q and also to $(Z_m\text{-}Z_r)$.

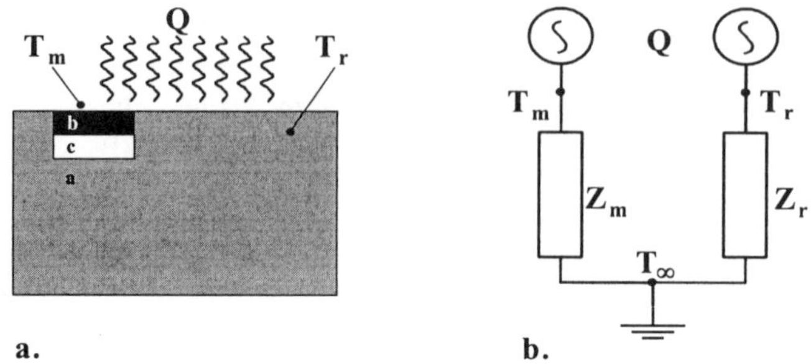

Figure 1: a) Schematic of a model showing a buried mine in the ground. b) Schematic of a simplified model to explain passive infrared imaging method using a constant heat flow and different heat impedances in the mine and the reference (no mine) sites. T_∞ is the temperature deep inside the ground.

Thus, in this method, it is the difference in heat conductivity of the mine and the soil surrounding it (at the mine level) that leads to a temperature contrast at the soil surface above the mine. Consider an initial state with a buried target at the same temperature as its surrounding soil and no initial temperature contrast at the surface. The onset of uniform heat flow Q from the atmosphere to the ground leads to a temperature contrast between the mine and soil surrounding it at the mine level due to different heat impedance's Z_m and Z_r as discussed above. There is a time delay between the onset of heat flow at the soil surface and the development of the temperature contrast at the mine level. This temperature contrast at the mine level is accompanied with the development of a temperature contrast at the soil surface [above the mine and the reference (no mine) region]. Enhanced surface heating, through infrared radiation, could increase the temperature contrast at the soil surface, and decrease the time required for the development of this contrast. However, infrared radiation cannot penetrate the soil. As such, it cannot interact with the mine *directly* to yield a temperature contrast at the mine level. Enhanced surface heating interacts with the mine only *indirectly* through the thermal conduction process. In contrast, microwaves penetrate a relatively long distance into the ground depending on the soil moisture content and interact *directly* with a mine buried near the surface and the soil surrounding it. The effect of this direct interaction can lead to a temperature contrast at the soil surface in a shorter time period than in an active IR (enhanced surface heating) imaging system.

When an electromagnetic wave travels from medium 1 to medium 2, a part of the incident radiation is reflected and the balance is transmitted into medium 2. For normal incidence, the reflection coefficient $\Gamma_\|$ for parallel polarization electric field is given by

$$\Gamma_\| = \frac{\sqrt{\dfrac{\varepsilon_1^*}{\varepsilon_2^*}} - 1}{\sqrt{\dfrac{\varepsilon_1^*}{\varepsilon_2^*}} + 1} \tag{2}$$

$$\varepsilon^* = \varepsilon' - j\varepsilon'' \tag{3}$$

ε^* is the complex permittivity and the suffix corresponds to the medium. Further, ε' is the relative dielectric constant and ε'' is the loss factor of the medium. The loss tangent is given by

$$\tan\delta = \frac{\varepsilon''}{\varepsilon'}. \tag{4}$$

The electromagnetic wave traveling in a medium is attenuated by a factor $e^{-2\alpha x}$ where x is the distance traveled in the medium. The attenuation factor α is given by

$$\alpha = \frac{\omega}{c}\left[\frac{\varepsilon'}{2}\left[\sqrt{1+\left(\frac{\varepsilon''}{\varepsilon'}\right)^2} - 1\right]\right]^{\frac{1}{2}} \tag{5}$$

where ω is the angular frequency of the electromagnetic radiation and c is the velocity of light. The power loss, p, (in dB/cm) of the electromagnetic wave due to the passage in the medium can be expressed as

$$p = -0.0868\ \frac{\omega}{c}\left[\frac{\varepsilon'}{2}\left[\sqrt{1+\left(\frac{\varepsilon''}{\varepsilon'}\right)^2} - 1\right]\right]^{\frac{1}{2}}. \tag{6}$$

The microwave power available for heating a sample buried at a depth x in the ground decreases exponentially with x and attenuation factor α. The increase in the temperature of the sample depends on the microwave energy absorbed by the sample and varies inversely with the sample's thermal heat capacity.

The propagation of microwaves in the soil and soil heating rates with microwaves have been studied by various workers.[12] Both the permittivity, ε', and loss factor, ε'', depend strongly on the moisture content of the soil, frequency of microwaves, and, to a lesser extent, the type of soil. In particular, at 2.45 GHz the loss factor, ε'', is quite low for dry sand and increases markedly with soil moisture content. Due to the difficulty in characterizing the soil type, it is best to measure the dielectric properties of the soil under consideration as a function of moisture content at the desired microwave frequency. Von Hippel [13] has given the dielectric properties at microwave frequencies of a variety of plastics and soils with different moisture contents.

The coupling of microwaves with the sample buried in the ground leads to a difference in temperature of the target and its surrounding soil, both at the target level and, after a time delay, at the soil surface level. At the same time, cooling mechanisms through conduction, radiation and convection set in to counter the development of these temperature contrasts.

3.1 THERMAL SIGNATURES AT THE SOIL SURFACE

There are two mechanisms that lead to a temperature contrast at the soil surface when microwaves interact with a buried mine. Consider the incident microwave energy transmitted into the soil at the soil/air interface (soil surface); a portion of the microwave beam propagates through the soil medium between the soil surface and the soil/mine interface. A portion of this beam will then be reflected from the soil/mine interface because of the variation in the dielectric constants. This reflected beam interferes with the incident beam in the region above the mine. The heating effect of the superimposed microwave fields at the **soil surface** provides a signature of the buried mine. This thermal signature will occur almost simultaneously with the microwave irradiation. The delay time is determined by the temperature resolution of the measurement instrument.

The second mechanism that leads to a thermal signature at the soil surface is due to the absorption of microwave energy by the mine. As discussed above, the buried mine and surrounding soil absorb microwave radiation at different rates, depending on their dielectric properties. This leads to a mine that is hotter or colder than the surrounding soil. The temperature difference is thermally conducted upwards to the surface above the mine and results in a thermal signature at the soil surface a short time after irradiation. The overall time evolution of the heating pattern at the soil surface will be a superposition of these two signatures. Depending on the depth of mine, the appearance of these signatures is generally well separated in time.

4. Numerical Simulations

Obtaining the experimental results for these trials is a laborious and difficult procedure. Measurements are complicated by;

- the need to maintain uniform moisture content over large areas of soil,
- uniform microwave illumination of the entire viewing surface,
- establishing sufficient microwave power density to obtain adequate heating rates,
- variations in convection cooling and heat diffusion rates with weather conditions,
- and precise positioning of the surrogate mines.

A series of experiments, often done over the course of several days, must be methodically compared and, unfortunately, this is rarely a simple procedure.

Numerical simulations have been used to complement and expedite the experimental process. Once an electromagnetic model of a mine surrogate is generated and the experimental validation of the simulation is complete, then variations in the IR surface image because of changing the mine depth (for example) can be easily done. The models are often overly simplistic, however they do establish trends that can later be confirmed experimentally. In addition, experimental results always involve multiple effects that cannot be isolated. For example, convection cooling will always corrupt the results of a measurement series, even if it is not one of the parameters under investigation. A simulation would allow convection cooling to be disabled so that a better understanding of individual parameters can be established.

4.1 NUMERICAL METHODS

The Finite-Difference Time Domain (FDTD) method was used for all simulations. The numerical model, which was designed by A. Louie and S. Kashyap of DREO, is depicted in Figure 2 and 3. The associated parameters are:

- Cell space (total number of Yee cells): 81 × 81 × 81
- Cell size: 5mm × 5mm × 5mm
- Frequency: 2.45 GHz
- Maximum reliable frequency (λ /10 rule): c / (10×5mm) = 6 GHz
- Number of cells modeling the buried object: 1864
- Number of time steps in an FDTD run: 1025
- Time step size (Courant limit): 5mm / ($\sqrt{3}\, c$) = 9.63 ps
- CPU time for an FDTD run on a DEC Alpha 1200: 33 min.

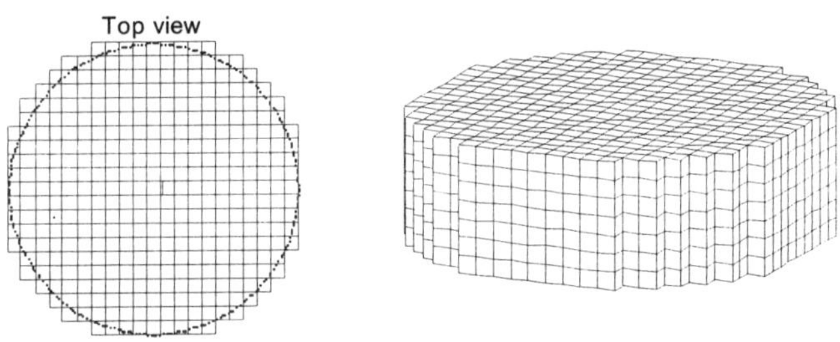

Figure 2: Numerical model of mine surrogate.

4.2 MODEL RESULTS

The FDTD simulations calculate an electric field distribution in and around the mine body. A typical example of the results is shown in Figure 4. The left-hand image is the electric field calculation with the soil/air interface at approximately z =13.5 cm. The right-hand images depict the conversion to power and then temperature (in the soil only) using the formulation.

$$\Delta temp = \frac{E^2 \Delta time}{\rho\, C_p}. \tag{7}$$

where E is the electric field, ρ is the density and Cp is the specific heat. Two problems exist in the comparison of numerical and experimental results. The first is the fact that the illumination used in the experiments was not uniform. A measurement of the power distribution for uniform soil with no mine surrogate can be used to compensate. This pattern can form the basis for a normalization of the results.

A second problem is the fact that the numerical model does not calculate thermal diffusion and convection. Every effort has been made to minimize these effects in the experimental results and, overall, the variations in the correlation to the numerical results are not significant. However, it would not be too difficult to incorporate diffusion and convection into the numerical models in the future.

4.3 VALIDATION TECHNIQUE AND RESULTS

Validation of the numerical model is a critical step in the evaluation process. As such, the cross-sectional temperature distribution in and around several surrogate mines was measured to establish the resolution and accuracy of the numerical results.

In order to establish this temperature distribution, infrared measurements of the heating patterns were made down to 50 mm below the soil surface. This was accomplished in a two-step process, as indicated in Figure 5. Initially, the exposure area was prepared by placing a series of wooden frames below the applicator. These rectangular frames are 30 cm by 45 cm, 1 cm thick and open in the middle, like an empty picture frame, to minimize the interaction with the microwave beam. The initial frame was laid in the soil surface at an appropriate distance from the applicator and the open central area was carefully filled with the same dry sand as was used in the surrounding pit. A layer of microwave plastic wrap (Saran Wrap^{TM}) was then placed on top of this frame to separate it from the next layer. This procedure was repeated for each frame until the assembly was completed.

The entire assembly was then exposed with a standard gain horn at a height of 26 cm. The exposure time was minimized to prevent convection and conduction cooling from excessively distorting the profile, but at the same time was long enough to provide adequate dynamic range for the IR camera (Agema Model #550). The IR image acquisition was started approximately one minute after completion of the microwave exposure, the minimum time required to remove the waveguide and position the camera. Immediately after photographing each layer, the frame and sand overburden were removed to expose the next layer. The plastic wrap facilitated a very quick and efficient removal of the soil. The entire imaging process took approximately two minutes. Calibration targets were also photographed to establish the dimensions of the images.

The experiment described above produced a series of images of the soil temperature at 1-cm increments, as shown in Figure 6. These images could then be digitized and processed in a matrix format. A cross-sectional cut could then be numerically established in any vertical plane. The image in Figure 6 has been interpolated to fill in the spacing between the layers.

f = 2.45 GHz	ε′	ε″	σ	Specific Heat [J/(gC)]
Polypropylene	2.25	0.00016	2.18e-5	1.8
phenolic	4.25	0.32	4.36e-2	1.8
dry sand	2.9	0.022	3.00e-3	0.79
Aluminum	---	---	PEC	

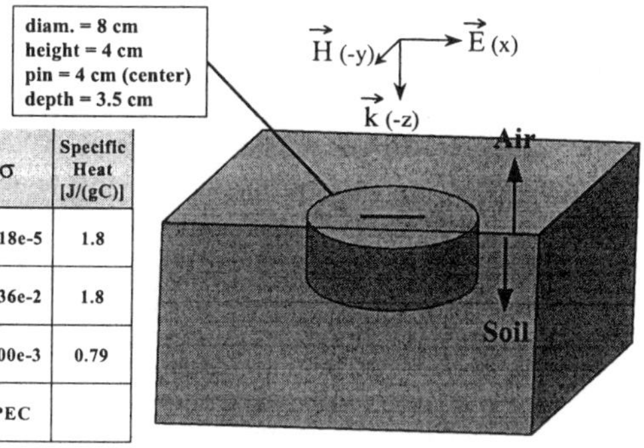

Figure 3: Dielectric properties of surrogate mine model.

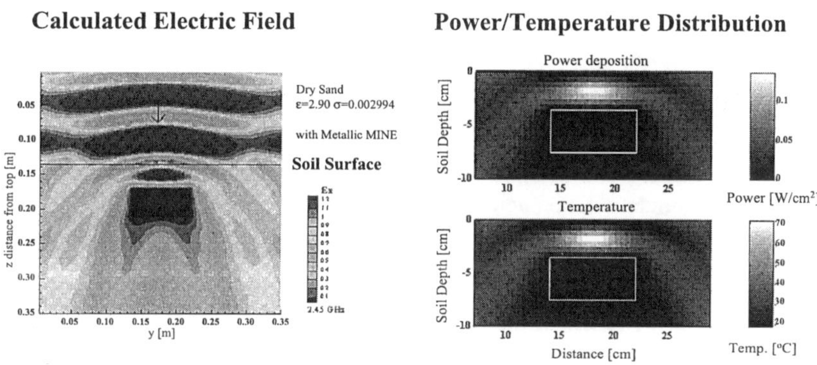

Figure 4: The left-hand image is a typical electric field calculation near a mine using uniform microwave illumination. The right-hand images depict the conversion to power and then temperature for comparison to experimental results.

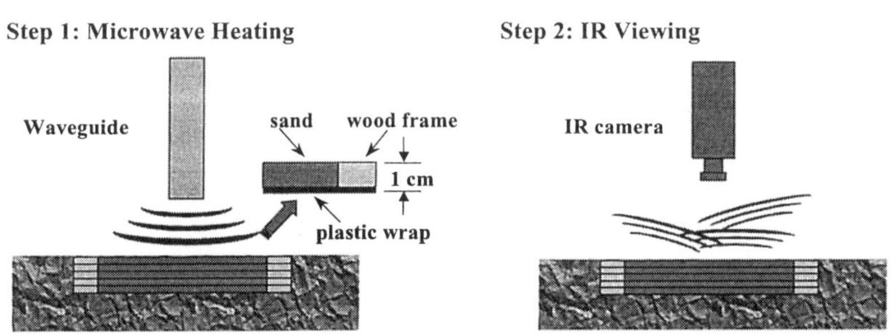

Figure 5: The temperature distribution around a mine surrogate was measured in a two-step process. Initially, the exposure area was prepared by placing a series of wooden frames below the applicator. The open area inside the frames was carefully filled with dry sand and a layer of plastic wrap was used to separate the layers. The entire assembly was then exposed to the microwaves and an IR camera was used to record the temperature distribution layer by layer. These images are then digitized and numerically cross-sectioned.

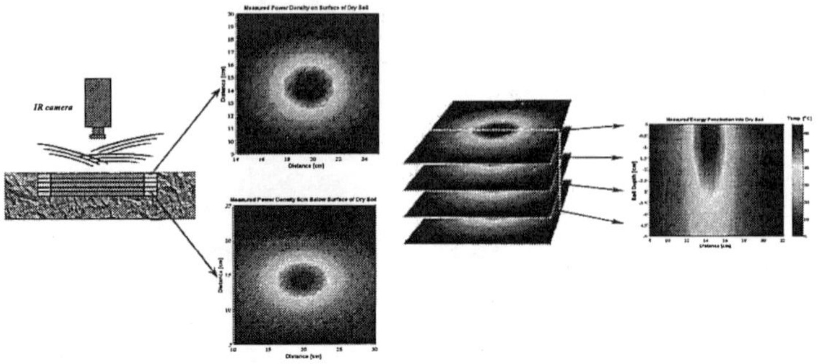

Figure 6: The series of images of the soil temperature could be digitized and stored in a matrix format, with each layer representing a given soil depth. A cross-section can then be numerically established in any vertical plane. The image above has been interpolated to fill in the spacing between the layers.

A comparison of modeling and measurement results is made in Figure 7. In this case, the temperature distribution in the soil surrounding a polypropylene mine surrogate was measured (left) and modeled using the FDTD technique (right). The mine surrogate was 8 cm in diameter, 4 cm thick and was buried 3.5 cm deep in very dry sand. In addition, the surrogate had a 4-cm metal pin place horizontally in the center of the mine body. In this particular simulation, the incident electric field was parallel to this pin.

Two cross-sectional cuts are shown in the simulation, The first (top) is a vertical plane in the center of the mine and perpendicular to the metal pin. The second (bottom) is a vertical plane in the center of the mine and parallel to the pin (i.e. perpendicular to the first plane). Comparisons are made for both sets of cut planes, however experimental data is not available in the layers occupied by the mine because of limitations in the measurement technique described above. In addition, the experimental data has been interpolated to enhance the visibility of the temperature gradients.

Excellent agreement is obtained both in the temperature distribution and peak magnitudes. It should be noted that the numerical model was calibrated by scaling the excitation electric field used in the simulation to that of the experiment, and that the temperature scale displayed has not been normalized in any other way. Similar agreement is obtained in Figure 8 where a metal surrogate has been used in place of the polypropylene. No other parameters have changed. Note that the plastic mine has a 'cold spot' at the soil surface and a 'hot spot' immediately above the mine. On the contrary, the metal mine has a 'hot spot' at or slightly below the soil surface. These variations are discussed in more detail in the Experimental Section on this paper.

Finally, Figure 9 and Figure 10 both represent the IR image of a polypropylene surrogate model at the soil surface. In Figure 9, the metal pin representing the fuse was parallel to the electric-field excitation. In Figure 10, the excitation was perpendicular to the pin. In both cases, the images comparing the measured (left) and modeled (right) results are in very good agreement. As in previous cases, the excitation used in the simulation was scaled to match that of the experiment. With the exception of the electric field orientation, all parameters remain the same in this series of images. There is clearly a very different response in the experimental results, which is accurately reflected in the numerical simulations. It would appear that the simulations provide a reasonable substitute for experimental results.

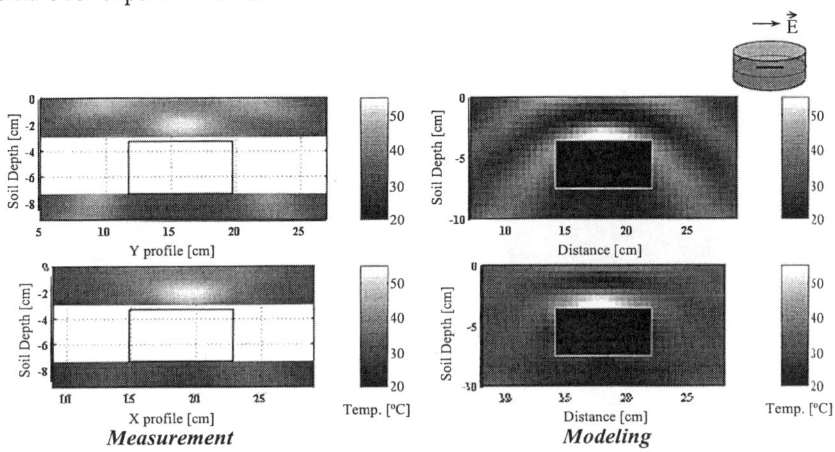

Figure 7: Validation of polypropylene surrogate model with metal pin representing fuse. The electric field excitation was parallel to the fuse. The top & bottom images represent two cross-sectional cuts through the same mine, one parallel and the other perpendicular to the pin, respectively. The excitation used in the simulation was scaled to match that of the experiment, however no other normalization is done. The calculated temperature distribution is in excellent agreement in both magnitude and shape.

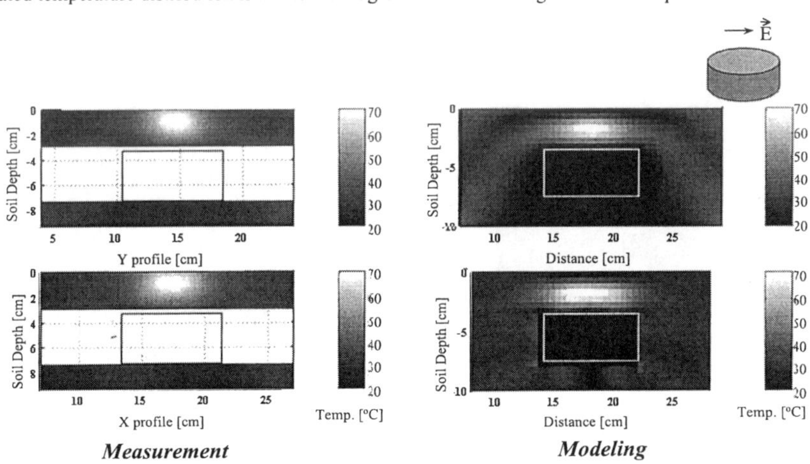

Figure 8: Validation of metal surrogate model. The top & bottom images represent two cross-sectional cuts through the same mine, as in Figure 7. The excitation used in the simulation was scaled to match that of the experiment, however no other normalization is done. The calculated temperature distribution is in excellent agreement in both magnitude and shape.

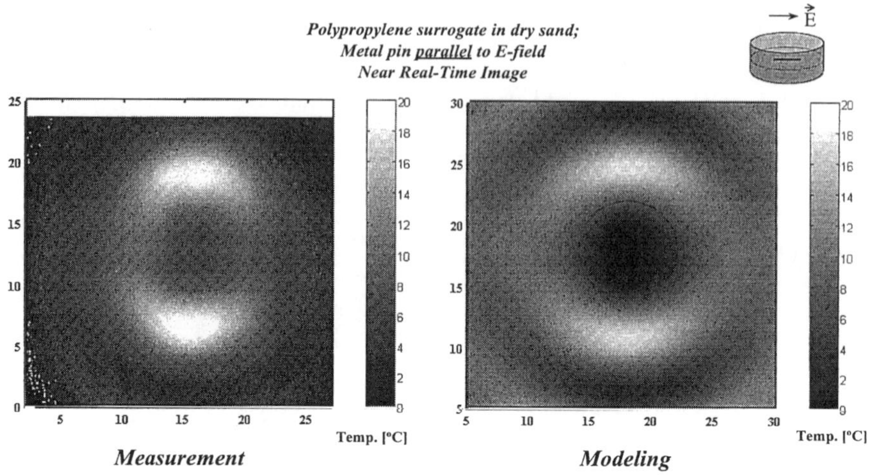

Figure 9: Validation of polypropylene surrogate model with metal pin representing fuse. The electric field excitation was *parallel* to the pin. The images compare the measured and modeled image of the mine at the soil surface. The excitation used in the simulation was scaled to match that of the experiment, however no other normalization is done. The calculated temperature distribution is in excellent agreement both in magnitude and shape.

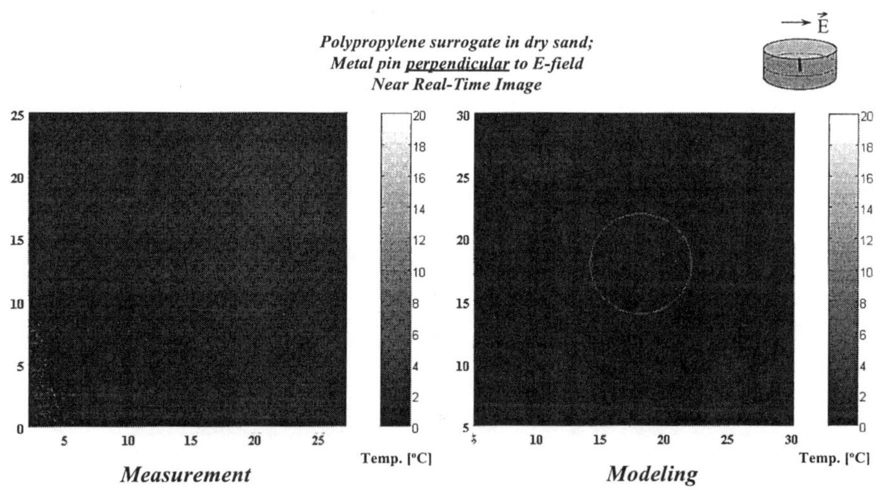

Figure 10: Validation of polypropylene surrogate model with metal pin representing fuse. In this case, the electric field excitation was *perpendicular* to the pin. All other parameters, including exposure power and time, remain the same as the above figure. The results are in good agreement and are drastically different from Figure 9.

5. Experimental Results

In the present work, the primary goal was to obtain accurate experimental data for a variety of scenarios under laboratory and field conditions to evaluate the suitability of this method for mine detection. A 5kW magnetron operating at 2.45 GHz was used as a source. Uniform illumination with a collimated microwave beam would be preferred for these experiments. However, for the present project, a waveguide (WR 284) with a standard gain horn antenna (EMCO 3160-03) was used to obtain a large cross-section beam. The targets were placed under the horn symmetric to its principal axis. The horn to soil surface distance could be changed by adding sections of waveguide. Most of the work was done with a horn to soil surface distance of 56 cm, although data was also taken with distances ranging from 30 to 88 cm. The microwave power output at the magnetron was varied from 1 to 5 kW. Typical time of microwave illumination ranged from 1 to 3 min. The microwave power density at the soil surface was estimated to be 1 to 3 W/cm^2. An 8-12 μm infrared camera provided infrared imagery of the soil surface covering the target.

Non-metallic mine surrogates consisting of discs (7.5 to 10.0 cm dia, 2.5 cm thick) of delrin, low density polyethylene and phenolic plastic were used. In addition, dummy mines without explosives and live mines with explosives but without fuses were also studied. Three types of mines were used. These are PMA 1, PMA 2, and PMA 3 and are shown in Figure 11. The dimensions of these mines are: PMA 1: length 140 mm, width 70 mm, height 30 mm, mass 400 g; PMA 2: dia 68 mm, height 61 mm, mass 135 g; PMA 3: dia 111mm, height 40 mm, mass 180 g. All three mines are minimum metal AP mines. The PMA 1 is rectangular in shape with a plastic body and a hinged top. The PMA 2 has a cylindrical plastic casing. The PMA 3 is a cylindrical in shape and is covered by a black rubber gasket.

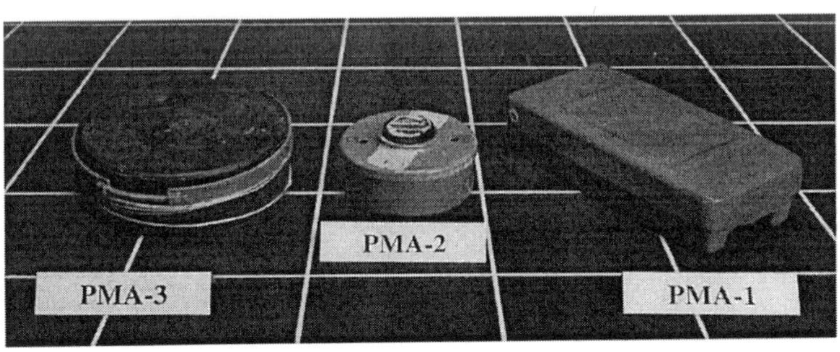

Dry and moist sand with moisture content ranging from ~ 0 to 20 % water content by mass were used for work in the laboratory. Laboratory experiments with mine surrogates and mines without explosives were also conducted with soil of low clay content and at various moisture levels. The depth of the top surface of the targets ranged from ~ 0.5 cm to ~ 5 cm, although most of the work was done with ~1 to 3 cm depth. In all laboratory experiments, special effort was made to ensure that the target and surrounding soil were initially at the same temperature and in an equilibrium state before microwave illumination. In the early stages of this project, most of the work was done with non-metallic (and metallic) mine surrogates inside an anechoic chamber. A 1 m tall plastic barrel full of sand was placed under the horn. The mine surrogates were placed symmetrically under the horn at desired depths in the sand. Initially, it was not possible to take real-time IR imagery and about 60 sec elapsed between the cessation of microwave illumination and IR imaging. Initially, automatic data acquisition was not possible with the camera and only Polaroid pictures were taken.

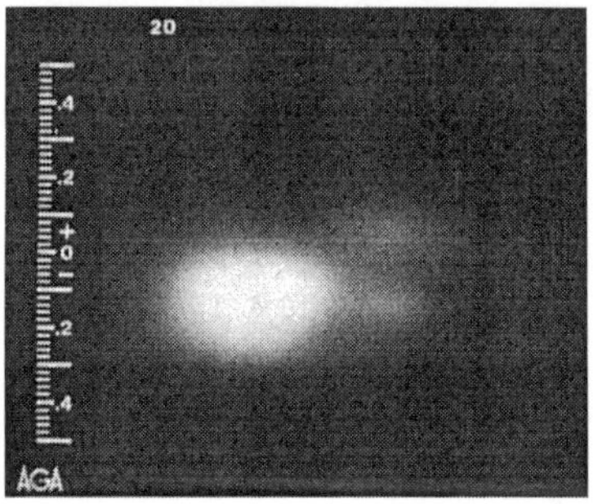

Figure 12: IR picture of the sand surface above three mine surrogates buried in dry sand after irradiating them with HPM

Figure 12 shows an IR picture of three mine surrogates consisting of a square phenolic (bottom left, 10x10 cm, 2.5cm thick, depth 1.1 cm), phenolic (bottom right 3.8 cm dia, 5 cm thick, depth 0.9 cm) and delrin (top right, 10 cm dia, 5 cm thick, 1.4 cm depth) discs buried in dry sand. The picture was taken 13 min. after irradiating the sand with HPM at 5 kW for 3 min. These IR signatures on the soil surface are due to the difference in microwave absorption by the mine surrogates and soil surrounding them at the mine level. The resulting temperature contrast at the mine level between the mines and the surrounding soil is conducted up to the soil surface and is recorded by the IR camera.

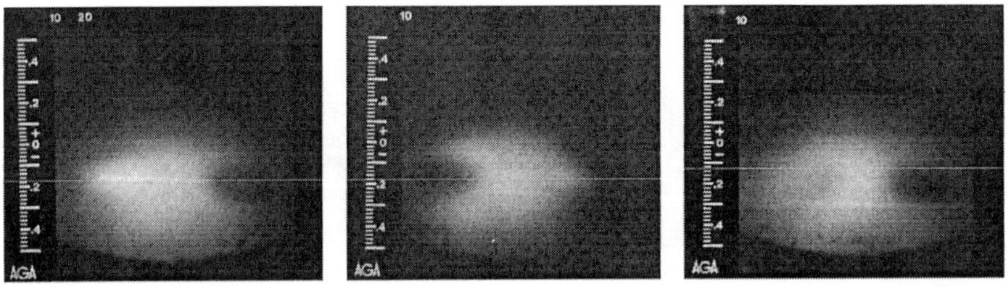

Figure 13a Figure 13b Figure 13c

Figure 13: IR pictures of soil surface showing the effect of reflection of microwaves from two mine surrogates buried at different depths in moist sand.

Thermal effects at the soil surface due to reflection of microwaves from mine surrogates buried in moist sand with 14% moisture content are shown in Figure 13. A microwave power level of 5 kW was used for 1 min duration. In Figure 13a two identical mine surrogates (delrin disc dia: 10 cm, thickness 2.5 cm) were buried at different depths (left target : 1.6 cm, right target: 1.0 cm). The central region in these pictures is still hot due to the heating of the soil by HPM. The two targets appear in Figure 13a as a hot spot on the left side and a cold spot on the right side against a general bright background. Although the two targets are of the same material, the path difference at the soil surface between the incident and reflected beams is different for these two targets due to the difference in their depths. This leads to a constructive interference at the surface for the left target and a destructive interference for the right target.

For confirmation of this explanation, the depth of these targets was changed in Figure 13b. The new depths are: (left target: 0.9 cm, right target: 1.5 cm). Note that the left signature changed from a hot signature in Figure 13a to a cold signature in Figure 13b. Similarly, the right signature changed from a cold signature to a hot signature in these pictures.

Additional confirmation of the reflected phenomenon is given by the results shown in Figure 13c. This figure depicts different thermal effects due to reflection of microwaves from two targets made of different materials which are buried at the same depth (~ 0.6 cm). The left target is an aluminum disc while the target at the right is a delrin disc. The aluminum target gives a hot signature while the delrin target gives a cold signature. The opposite polarity of the signatures from these two targets at the same depth is due to the phase reversal of the microwave field upon reflection at a metallic surface. Optimal IR interference patterns are obtained in a uniform medium, as was the case in Figure 13. In a non-uniform medium, random scattering of the microwaves would tend to scatter the reflected signal from the target and degrade the interference pattern.

One means of eliminating the variations in the soil medium is to perform the experiment depicted in Figure 13 numerically. Figure 14 represents a simulation of the variation in IR surface patterns with respect to surrogate burial depth. The top row depicts a metal surrogate mine buried at 2, 4 and 6 cm. The lower row is a Bakelite surrogate buried at the same depths. Two observations can be made: the first is that the same mine can generted hot or cold signitures depending on the burial depth. The second is that at the same burial depth, two mine types can generate very different signitures. This is consistent with the observed experimental results.

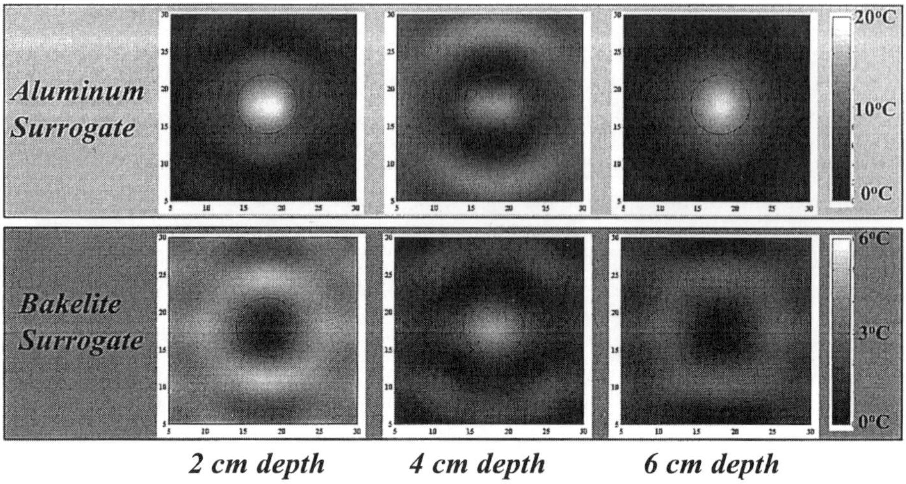

Figure 14: Numerical simulation of variation in IR surface pattern with surrogate depth. The top row depicts a metal surrogate mine buried at 2, 4 and 6 cm depths. The lower row is a Bakelite surrogate buried at the same depths. Two observations can be made: the first is that the same mine can generted hot or cold signitures depending on the burial depth. The second is that at the same burial depth, two mine types can generate very different signatures.

For real-time HPM/IR imaging and better controlled laboratory experiments, a 2 m x 2 m x 0.6 m deep pit was filled with the same construction sand as was used in the previous experiments; see Figure 15. Only the moisture content of the sand immediately below the horn (applicator) was manipulated for the experiments. For the majority of work, an insulated wooden structure was constructed around the pit to minimize variations in the environment above the soil surface. Real-time IR imagery was acquired during and after HPM irradiation using an 8-12μm Agema Thermovision 880 IR camera. Digital image processing of the IR imagery was done to obtain the target signatures.

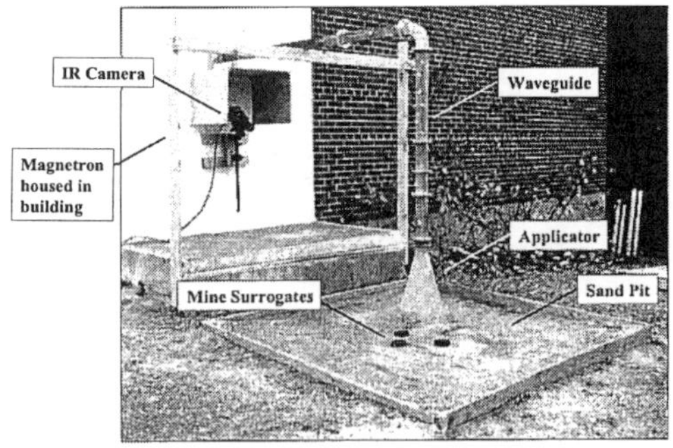

Figure 15: Experimental set-up for real-time HPM/IR imaging outside the anechoic chamber. The height of the horn could be adjusted.

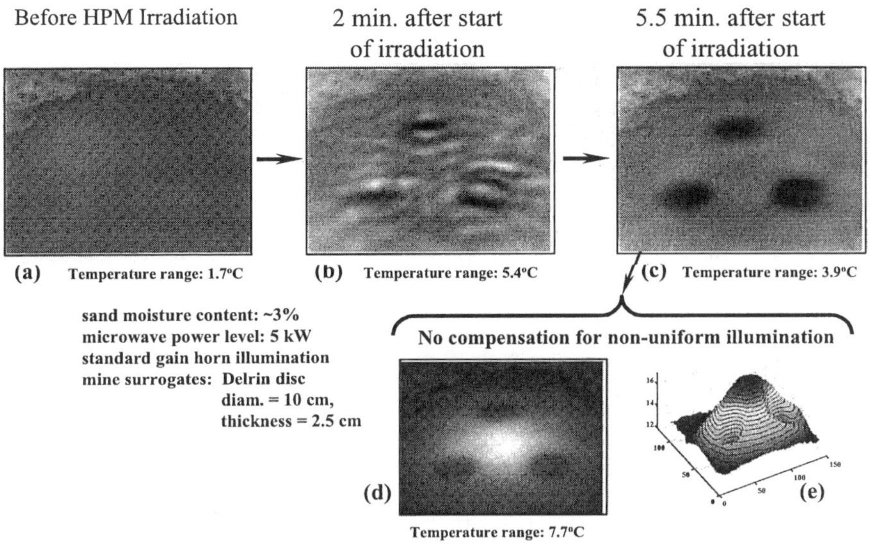

Figure 16: IR pictures of the sand surface over mine surrogates buried in moist sand at different depths. The temperature range of the data is shown in each picture. a: Before HPM irradiation; b: 2 min. after irradiation; c: 5.5 min after HPM irradiation; d: 5.5 min after irradiation but without any compensation for non-uniform HPM illumination; e: A 3D representation of (d) showing IR intensity across the image. HPM spot intensity has been removed from Figure 16b and c. Figure 16c and d are the same except that the spot has been removed in (c).

Figure 16 depicts IR pictures of a moist sand surface (moisture content: 7.7% by weight) covering 3 mine surrogates buried at different depths. The surrogates are Delrin discs, 10 cm diameter and 2.5 cm thick. The depth of the targets: (Right bottom: 0.5 cm, Left bottom: 0.8 cm, Top: 1.0 cm). The targets were heated by HPM at 5 kW for 1.5 min. The horn to sand surface distance was 56 cm. IR pictures of moist sand surface were taken before, during and after HPM irradiation. The top portion in these pictures has not received any significant HPM radiation. In Figs. 16b & 16c, the intensity has been compensated for non-uniform microwave illumination.

Figure 16a shows almost no IR signature of the mine surrogates at the soil surface before HPM irradiation. Figure 16b shows the IR signatures at the surface due to reflections from the mine surrogate as discussed previously. The three mine surrogates could be seen clearly, although the pattern of reflection does not show the complete target. Figure 16c gives very clear thermal signatures on the sand surface for the three mine surrogates. These signatures are due to the difference in microwave absorption by the targets and the sand at the target level surrounding the targets. This thermal contrast at the target level is then conducted upwards to the soil surface and appears at the soil surface sometime after HPM irradiation. It should be noted that the targets in Figure 16c and

16d appear colder than the background in contrast to the results shown for dry sand. These results would be expected from a comparison of the loss factor of sand, moist sand, and these targets. Figure 16d was taken 5.5 min after microwave radiation was turned off. It includes the effects due to non-uniform heating of the sand surface resulting from the microwave beam pattern under the horn. Although the targets are apparent in this figure, numerical compensation for the non-uniform HPM irradiation will significantly improve target detection capability. The temperature at any point or average temperature for any small region in this picture could be determined. A 3D plot of intensity variation across Figure 16d is given in Figure 16e, which clearly shows the presence of mine surrogates.

Extensive data was taken with other types of plastics targets and soils containing clay with varying moisture content. Data was also taken in the laboratory with PMA 1, PMA 2 and PMA 3 mines without explosives. PMA 1 was filled with paraffin to duplicate the microwave properties of explosives. This data is not included here due to lack of space. However, it should be mentioned that almost in all cases, the targets were detectable after HPM irradiation through thermal signatures on the soil surface, either through the reflected signal from the mine, or the differential absorption by the mine as compared to the soil surrounding it or both.

Figure 17: Experimental arrangement used in field trials at DRES for the detection of real mines using the HPM/IR method.

Results of field trials conducted at the Defence Research Establishment Suffield (DRES), Canada in April 1997 are described next. PMA 1, PMA 2 and PMA 3 mines with explosives but without fuses were used. The experimental arrangement is shown in Figure 17. Using a trailer, a horn could be moved directly over the mines buried in the ground. The magnetron and IR camera used for this work at DREO were also used in these trials. Real-time IR imagery was collected before, during and after microwave irradiation.

Figure 18: Standard photographs showing PMA 1 and PMA 3 buried in soil without turf (left) and soil with turf (right). The soil above the mines had been removed to take this photograph.

The mines were buried in the ground four days prior to these experiments by personnel experienced in this work. There was no rain during this period. The climate was sunny, cold, windy and dry with an air temperature of 6°C. The PMA 1 and PMA 2 were buried in the ground from which turf was already removed. The size of loose disturbed soil without turf, covering each of these mines, was 22 x 22 cm (see Figure 18). In addition, the PMA 2 and PMA 3 were buried in soil covered with turf that had dried out over the winter season. A 12 x 12 cm turf patch covered each of these mines (see Figure 18).

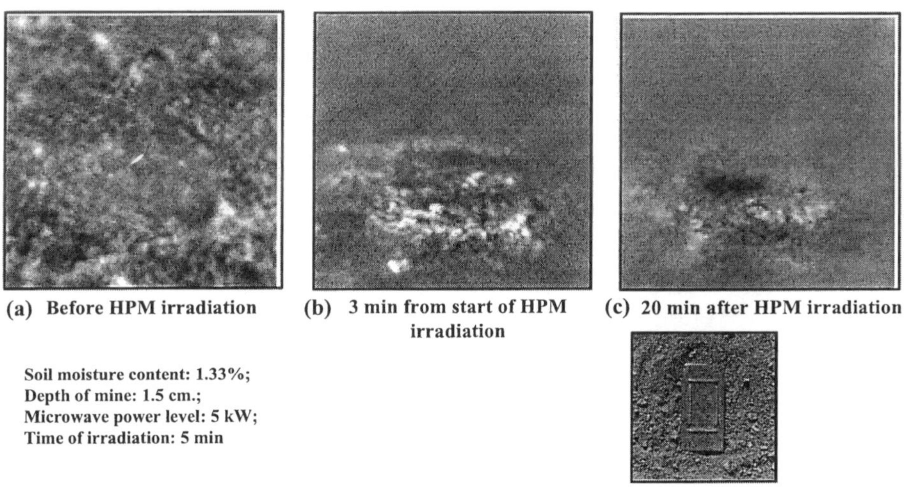

Figure 19: IR pictures of loose soil covering PMA 1 containing explosives. a: Before HPM irradiation; b: 3 min from start of HPM irradiation; c: 20 min after HPM irradiation. Soil moisture content: 1.33%; Depth of mine: 1.5 cm.; Microwave power level: 5 kW; Time of irradiation: 5 min. The mine is visible due to microwave absorption by the mine as shown in (c). Results due to reflection of microwaves by the mine are ambiguous (b). Enhanced thermal features in this picture could be related to the mine.

Figure 19 is an IR picture of the loose soil surface, without any grass, covering the PMA 1 containing explosives, but without a fuse, buried at a depth of 1.5 cm. The loose soil patch is not distinguishable in the IR image before microwave irradiation (Figure 19a). Figure 19b depicts the thermal effect on the soil surface due to reflection from the mine. This figure shows that the mine is not clearly distinguishable. However, enhanced thermal features on the surface, especially dark and bright bands over the mine region could be noted. In a similar experiment in the laboratory but with 6% moisture content, dark and bright bands or patches were seen over the mine region due to reflections from the mine. Figure 19c, taken 20 min after irradiation, depicts the thermal signature at the soil surface due to a difference in microwave absorption by the mine and the surrounding soil. PMA 1 is clearly distinguishable from the background in Figure 19c.

Note that the loose soil covering the mine is not visible in IR before HPM irradiation (Figure 19a), but can be more easily demarcated after HPM irradiation (Figure 19b). The square shape of this loose soil patch can also be identified.

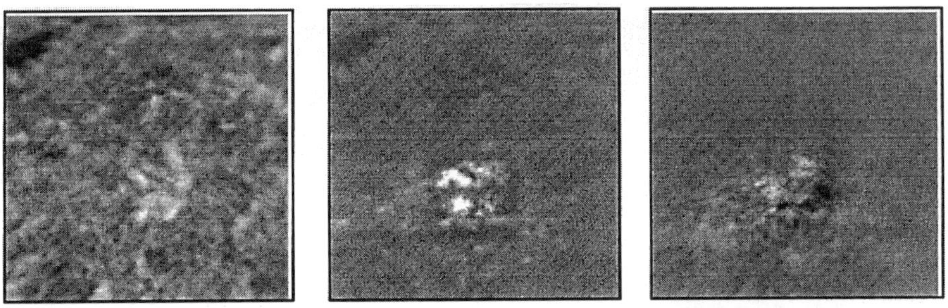

(a) Before HPM irradiation **(b) 2.5 min from start of HPM irradiation** **(c) 14 min after HPM irradiation**

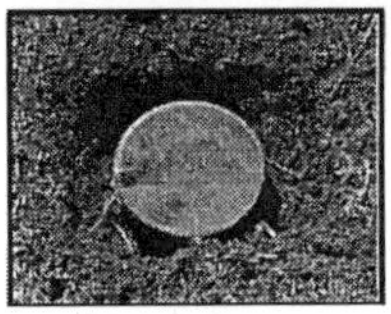

Soil moisture content: 2.8%;
Depth of mine: 2.5 cm.;
Microwave power level: 5 kW;
Time of irradiation: 5 min

Figure 20: IR picture of soil turf covering PMA 3 (with explosives). a: Before HPM irradiation; b: 2.5min from start of HPM irradiation; c: 14 min after HPM irradiation. Soil moisture content: 2.8%; Depth of mine: 2.5 cm.; Microwave power level: 5 kW; Time of irradiation: 5 min

Similar results are shown for PMA 3 in Figure 20. The mine is buried in a smaller patch of turf, 12 x12 cm in size. Enhanced thermal structure, which may serve as a mine signature, can be seen over the mine location in Figure 20b. Comparing Figure 20a and b, it is also clear that it is much easier to pinpoint the disturbed turf area used to bury the mine with this method than through passive IR imaging. The square shape of this cut turf piece can be identified. Compare the larger size of the disturbed area in Figure 19b to that in Figure 20. For the PMA 3, a thermal signature, due to absorption of microwaves by the mine, is not seen. The effects due to reflection appear to mask the effects due to absorption. In contrast, thermal signals due to both the reflection and absorption by PMA 1, PMA 2 and PMA 3 were seen with mine depths ranging from 1 to 2 cm with a soil moisture content of ~ 6%.

6. Conclusions

The results presented in this paper confirm that the HPM/IR method can provide a powerful remote-sensing technique for mine detection. This method is complementary to passive IR mine detection techniques and can detect mines under cloudy and wet conditions. The mine signatures arise due to both reflection and absorption by the mine and appear as thermal signatures at the soil surface above the mine. The measurements presented are clearly very preliminary. Much further work is needed to fully understand the strength and limitations of this method, especially under field conditions. The false alarm rate, detection sensitivity and detection rate have to be determined and would ideally be better than those available from existing methods. This method can be used as a pre-confirmatory method for precise localization of a mine from a stand-off distance.

7. Acknowledgements

The authors would like to thank D.Liang, J.R. Evans, J.E. McFee, D. Bergeron, C. Gardner, S. Kashyap, A. Louie, D. Duermeyer for their support of this project.

8. References

1. www.care.org:80 world wide web-site
2. J.E. McFee, Y. Das, A. Carruthers, S. Murray, P. Gallagher, and G. Briosi, "**CRAD Countermine R&D Study - Final Report**", Defence Research Establishment Suffield Special Publication 174, April 1994.
3. J.-R. Simard, "**Theoretical and experimental characterizations of the IR technology for the detection of low-metal and non-metallic buried landmines**" DREV-R 9615, Defence Research Establishment Valcartier, Quebec, Canada, March 1997.
4. T. Cousins, T.A. Zones, J.R. Brisson, J.E. McFee, T.J. Jamieson, E.J. Waller, F.J. LeMay, H. Ing, E.T.H. Clifford, and E.B. Selkirk, "**The development of a thermal neutron activation (TNA) system as a confirmatory non-metallic land mine detector**", Presented at MARC IV, Kona, Hawaii, April 1997.
5. P. Li, A. Maad, F. Moshary, M.F. Arend, and S. Ahmed, "**Infrared imaging of buried objects by thermal step-function excitations**", Appl. Optics, **34**, pp. 5809-5816, 1995.
6. K. Fueloep and R. Bird, "**The Detection of Landmines using Infrared Thermal Imaging and Microwave Irradiation**", DSTO Report DSTO-TR-0402.
7. L.J. Carter, G.H. Bryant, M. Le Fevre, and W.C. Wong, "**Moisture and landmine detection**", Proc. of the Conference on Detection of Abandoned Land Mines, IEE Conference Publication No. 431, pp. 83-87, 1996.
8. J. Hermann and I. Chant, "**Microwave Enhancement of Thermal Landmine Signitures**", SPIE Vol. 3710, April 1999.
9. C.A. DiMarzio, C.M. Rappaport, L. Wen, G.O. Sauermann and H.E. Scott, "**Microwave-Enhanced Infrared Thermography**", SPIE Vol. 3710, April 1999.

10. S.M. Khanna, F. Paquet, R. Apps and J.S. Seregelyi, **"New hybrid remote sensing method using HPM illumination/IR detection for mine detection"**, Proc. SPIE, Vol. 3392, p. 1111-21, 1998.
11. J.S. Seregelyi, S. Kashyap, A. Louie and S.M. Khanna, **"HPM/IR Detection of Land Mines: Measurement and Modeling of Surrogate Mines"**, EUROEM 2000, May 2000.
12. J. S. Seregelyi, R. Apps, M. Boyle and C. Gardner, "**Microwave Heating of Soil**", DREO Report, Defence Research Establishment Ottawa, Ottawa, Canada, 1998.
13. R. Von Hippel, ***Dielectric Materials and Applications***, pp. 314-327, M.I.T. Press, Aug. 1966.

THE OPERATIONAL EXPERIENCE AND PROSPECT OF DEVELOPMENT OF THE DEVICES FOR DETECTION OF EXPLOSIVE SUBSTANCES BY THE THERMAL NEUTRON RADIATION ANALYSIS METHOD

OLSHANSKY Y.I., FILIPPOV S.G.
S&TC RATEC, Ltd. Okjabrskaj quay, h 44/2, St. Petersburg, 193079, box 84, Russia
KYZYUROV V.S., LAYKIN A.I.
Krylov Shipbuilding Research Institute, 44, Moskovskoe shosse, St. Petersburg, 196158, Russia

1. Introduction

The problem of explosive detection has become more acute over the last 10-15 years. Many attempts have been made to use thermal neutron activation analysis (TNA) to develop multi-purpose devices. The devices built upon that principle detect a greater (compared to the standard one) level of nitrogen concentration, which is common to all widely used explosive substances. The Explosive Detection Devices (EDD) use a thermal neutron flux to irradiate the inspected objects. When nitrogen nuclei capture thermal neutrons, characteristic gamma rays are emitted with energy 10.8 MeV. Monitoring of gamma radiation enables one to determine nitrogen concentration in the examined objects.

The advantage of neutron radiation explosive detection devices compared to the commonly used X-ray machines is based on the fact that detection of explosive substances is done automatically without an operator usually needed for the visual identification of hidden explosives.

EDD detect explosive substances that can not be found by commonly used X-ray machines. Such explosive substances include plastic bombs hidden in cameras, radios and tape-recorders, or disguised as personal items such as soap, batteries, briefcase and luggage hardware, etc.

The attempts to escape the neutron flux of the device using the materials capable of absorbing thermal neutrons can be detected by monitoring any distortions in the thermal neutron field.

Unlike vapor and particle sniffers the Explosive Detection Devices are capable of detecting even sealed explosive substances.

In spite of the attractiveness and simplicity of the TNA method the actual development of a working EDD device is complicated by a number of technical and engineering problems. Because of these complications, as far as we know, so far only two companies have been able to built working EDD systems. One of these companies is SAIC based in St. Clara, CA, the other one, RATEC, is in St. Petersburg, Russia.

The problems arising before the developers of devices for detection of explosive substances that use TNA method, are common for different applications of this method, including use for detection of mines.

H. Schubert and A. Kuznetsov (eds.), Detection of Explosives and Landmines, 125–131.

2. Problems of the device for detection of explosive substances on the basis of TNA method

During the last several years, S&TC RATEC has developed a number of the prototypes and samples of devices for detection of explosive substances for various applications, including the following:

Two modifications of devices for detection of explosive substances (ES) in carry-on luggage of air passengers and in other objects (EDS-21 and EDS-2101) with following basic characteristics:

- Minimum mass of the ES (nitrogen concentration not less than 25% by weight) - not more than 400 grams for any items, and 100 g for the items with a small amount of nitrogen (such as radio electronics equipment, etc.) the probability of detection is 95%, the probability of false alarms is 1%.
- The internal sizes of the inspection chamber is 500-500-250 mm (depth/length/width)
- Inspection time - 15-30 seconds.

Devices for detection of explosive substances in the items of mail and in other subjects (EDS-3101) with following basic characteristics:

- Minimum mass of the ES (nitrogen concentration not less than 25% by weight) - not more than 100 grams with the analysis time not more than 100 seconds, with the probability of detection at 95%, the probability of false alarms of 1%.
- The average detection time of the ES of not more than 150 g is 20 seconds.
- The internal sizes of the inspection chamber is 120-300-360 mm (depth/length/width)

Preliminary sample of the device EDS-5101A for detection of explosive substances intended for the working in combined system with X-ray machine and operating for the best detection of explosive substances the information on site and weight of a suspicious subject in checked object gained from X-ray machine.

Fig. 1 The prototype EDS-21 at experimental operation in an airport Pulkovo-1, St.-Petersburg

Fig. 2 The prototype EDS-2101 at laboratory , RATEC, St.-Petersburg

Preliminary sample of the device for simultaneous detection of explosive and fissionable nuclear substances with use as a source of neutrons (D,D) neutrons generator together with Krylov Shipbuilding Research Institute within the framework of activities on the project # 1509 of ISTC.

The prototype of the device EDS-21 was during several years in experimental operation in an airport Pulkovo-1, St.-Petersburg on an actual flow of carry-on luggage air passengers.

The prototypes of the devices EDS-2101 and EDS-3101 subjected to intensive tests in laboratory of the enterprise

The general view of the prototypes of these devices is shown in figures 1,2 and 3.

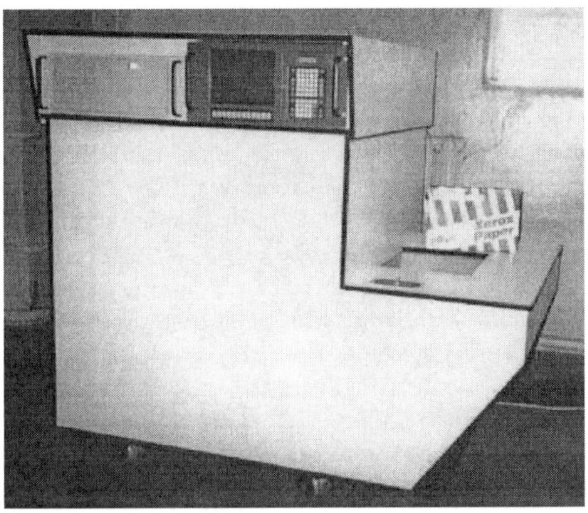

Fig. 3 The prototype EDS-3101 at laboratory , RATEC, St.-Petersburg

Let's look at some of the problems impeding the development of Explosive Detection Devices and possible ways of their solution on the basis of experience of development and tests of these prototypes.

First, let's consider the technical problems. One of the main problems here is the presence of nitrogen in commonly used objects and materials that can be placed inside the examined objects, such as, food products with high protein, nylon, wool, polyurethane and other materials. This can be a cause for a high number of false alarms while examining passengers' check-in and carry-on luggage.

We conducted a study to determine the probability of various quantities of nitrogen present in personal belongings and items of air-travel passengers' carry-on luggage. As a result we have come up with a formula for determining nitrogen content:

$$P\,(M_N) > EXP\,[-M_N / (Ma \cdot Vc)]$$

$P\,(M_N)$ - probability of nitrogen present with a mass greater than M_N;

Ma - average statistical mass of nitrogen present in the "carry-on" luggage;

Vc - examined volume of the "carry-on" luggage

All of the data was collected over a one-year trial period at Pulkovo-1 airport in St. Petersburg using EDS-21, an explosive detection device for air passengers' carry-on luggage. These data correspond to the similar data collected by SAIC on a trial basis while testing a device for explosives detection developed by that company [1,2].

Using the above-given formula and trying to achieve an acceptable false alarm probability rate (Pfa < 0.01), it was determined that the examined volume should not exceed several liters, which is almost impossible from the technical point of view for any luggage greater than several dozens of liters. A high number of false alarms from the devices made it impossible to introduce SAIC systems at the airports. This prompted us to come up with a design for a device with a smaller chamber for the inspection of "carry-on" and hand luggage of the air-travel passengers

The above-mentioned technical problem can be solved in different ways:

the inspection of the luggage with a known small level of nitrogen concentration: such as, electrical, photo-and computer equipment, which at the same time allows a very efficient disguise of the explosives.

- utilization of the X-ray machines for the preliminary identification of the suspicious areas and their subsequent inspection with an EDD device
- maximum narrowing of the controlled volume by the narrowing of the neutron field and the narrowing of the sensitivity curve of the sensors

Our calculations show that it is possible to develop an explosive detection device to inspect air-travel "carry-on" baggage identified by an X-ray machine as suspicious, and exposing only the problem areas to the neutron radiation and subsequent processing of the data collected from the gamma-detectors.

EDDs with small inspection chambers can be used to examine visitors' and staff personal bags with a small number of nitrogen-containing items. Such EDDs can be installed in the government office buildings and high-security facilities such as nuclear installations. For the use at nuclear facilities the devices can be upgraded to detect any unauthorized transfers of nuclear materials.

In addition to the technical difficulties associated with a large number of false alarms, the development of the Explosive Detection Devices using the Neutron Radiation Analysis is complicated by a number of engineering/technological problems. The most common technological problems are:

- the creation of an equally distributed maximum high thermal neutron field in the working chamber of the device
- the determination of the "background" count rate (in the absence of nitrogen in the controlled object) when nitrogen captures the neutrons in the field of gamma- (quants) radiation
- the development of high-speed electronics with greater efficiency rejection of overlapping pulses for the recording of gamma radiation;
- the procedure of the gamma-spectrometer calibration and the algorithm of the decision-making process.

3. The proposals on directions of activities for the decision of problems

Now let's look at each of these problems in detail. The creation of an equal density thermal neutron flux in the inspection chamber is necessary in order to receive equal responses from the explosive substances no matter where they are placed in the controlled item. In addition, this has to be achieved with a minimum neutron source and the smallest possible size and weight parameters of the device to satisfy health and biological safety requirements.

A successful solution of this problem becomes extremely important in the development of smaller devices, in which graphite cannot be used as absorber. To solve this problem we used the concept of neutron projector, which allows to increase the thermal neutron flux by 1.4 times compared to the regular absorber (a ball of hydrogen-containing medium with a neutron source in the middle) while keeping the size of the device small. In addition, this design of the thermal neutron source reduces the movement of the neutron flux away from the inspection chamber, which, in turn, provides better biological safety. Special placement of the neutron source (Cf-252) inside the projector and the use of the bismuth shielding allows to lower gamma radiation from the source which is picked up by the sensors

Based on the design and experimental data we developed projectors for each EDD model. For example, for EDS-3101, an EDD system for mail inspections, we received the following parameters for the creation of the thermal neutron field: coefficient of heterogeneity of the thermal neutron field density in the working chamber - 0,8, ; the source to thermal neutron conversion rate - $K = 9 \cdot 10^{-4}$ [1/ cm^2]

The correct estimation of the background count rate of each gamma spectrometry trail in the radiation field of the captured nitrogen is necessary for the successful operation the device. The presence of an explosive substance is determined by an increased count rate of the energy spectrum in of the captured nitrogen gamma radiation, i.e. 10.8 MeV. The benchmark value of the count rate, the increase of which identifies greater that normal nitrogen concentration, is calculated using the background count rate of that area.

Generally, the background count rate depends on the size and composition of the inspected object, i.e. each item has a specific count rate in that particular area without any nitrogen present in it. Thus, for the detection of explosive substances we need to know the background count rate for each item. In the past, the detection algorithm used for calculations the background count rate of the empty chamber or of some sample baggage, which caused uncertainties for the items, which contained substances that distorted the neutron flux. This problem was partially fixed by using a benchmark bias, which was taken by measuring neutron fluxes in different parts of the inspection chamber. However, we were not able to come with a single bias for all cases, as there is no universal correspondence between the neutron fluxes and the background count rate.

Our study showed correlation between the energy range of 3 - 5 MeV and the working section of the device with energy of 10 - 11 MeV. This correlation allows us to calculate the background count rate in the working section with a high degree of reliability using the measured range of the captured radiation from the examined object. This correlation is observed for a wide variety of substances, which can be found inside the inspected object, and for a large section of the working chamber.

To increase the effectiveness and efficiency of the explosion detection device the data processing program uses several capable computing algorithms.

The sequence analysis algorithm.

Time is one of the most important parameters, since it determines the throughput of the device. For the devices that use nuclear-physical methods of detection the time attribute is closely associated with the reliability and accuracy of detection. On the other hand, the time needed to gather the necessary information to make the right decision also depends on the inspected item. If the item contains a large amount of the sought material, the available collection time will be less. This fact prompted us to use Wald's algorithm of the reduced sequence analysis. This algorithm minimizes the time needed to make the decision and makes it dependent on the amount of the explosive substance in the examined object. That is, the time needed to make the decision is always different, and the more explosive material you have in the item the less time is required for its detection. The maximum time for detection is needed only for the smallest amounts of explosives. This method has a dramatic time advantage over the fixed inspection time algorithm.

The partial inspection algorithm.

A special algorithm was needed for the inspection of the items containing nitrogen not associated with explosives. This algorithm takes into account possible nitrogen radiation from non-suspicious objects. This method finds its practical application when an EDD device is used in conjunction with an X-ray machine that identified suspicious objects in some parts of the inspected baggage, and can have harmless nitrogen-containing objects in other parts of the luggage. To solve this particular problem we used the sensitivity matrix algorithm. The matrix shows how sensitive each sensor of the device is towards the explosive substances. The algorithm determines the presence of ES based on the weighted sum of values (sensor data are multiplied by the weight coefficients, that are proportionate to the diagonal elements of the matrix). The algorithm has an additional parameter that eliminates false alarms in the case of large quantities of nitrogen in the non-examined sections of the controlled objects.

The two-calibration algorithm.

Good calibration of the gamma spectrum is important for the selection of the working range for the finding of the captured radiation. The main difficulty of the calibration is in the small amount of gamma data received from the scintillating detectors. To increase the reliability of the performance the devices use two independent methods of the gamma spectrum calibration: the first one, using the location of the most typical photo peaks for the empty inspection chambers (e.g., the hydrogen photo peak); the second one - using previously measured points of the integral spectrum. The final calibration is done after the compared results of the two calibrations show no relevant differences. The results from the two calibrations can be used as an additional maintenance test of the equipment.

To detect and record gamma rays with high-efficiency rejection of overlapping pulses the neutron radiation analysis devices need to use high-performance electronics. We have developed an analog-digital converter, and a complete gamma spectrometry path, which allows to work with the loads of $5 \cdot 10^5$ pulse/sec, as well as an efficient rejecter of overlapping pulses for the scintillating spectrometers. The rejecter selects the overlapping pulses in the case of the disagreement with more than $6 \cdot 10^{-8}$ sec.

The in-service experience of installation UVP-21 in an airport Pulkovo-1 also has shown, that the reason of originating of false alarms in the device of a similar type is the presence in luggage of the air passengers of materials containing iron, chlorine, chromium, nickel and others, which under action of thermal neutrons radiate a gamma quanta with power close to power of radiation of nitrogen. The pulses from radiation of these elements can be received for pulses from radiation of nitrogen that can call a false alarm.

It is possible to exclude these false alarms with the help of special algorithm, which is taking into account magnifying of quantity of pulses in the field of 8-10 MeV, in which partially there is a radiation of these hindering materials. The experiments have shown, that such algorithm is implemented and results in practically full disappearance of false alarms, which are obliged to presence of the above-mentioned hindering materials.

Using all of the described approaches we develop now prototypes of the devices for simultaneous detection of explosive and fissile nuclear substances and special device EDS-5101 . First prototype (project #1509 ISTC) can be installed at different nuclear facilities to prevent any terrorist acts or unauthorized transfers of nuclear materials. To detect nuclear materials (U-235, Pu-239) these devices monitor delayed neutrons that are emitted during the fission process when they are bombarded by thermal neutrons and natural gamma radiation nuclear materials. This ensures a high level of detection reliability even of the nuclear materials that are being taken out in small shielded containers.

The high-sensitive exterior detectors of gamma of radiation and detectors of neutrons inside the chamber of the prototype are stipulated by its construction for a detection of natural gamma radiation and delay neutrons from nuclear materials.

References:

Trial Protocol of system EDS-21 in airport "Pulkovo", St. Petrsburg, January 22, 1993

Tsashi Gozany. Neutron Based Non Instructive Techniques. Obninsk Sumposium on Nuclear Physics Methods for Detection of Smuggled Explosive and Nuclear Materials. 1995.

Characterization of chemical sensors for the detection of explosives – development of standards.

Gerhard Holl

Bundeswehr Research Institute for Materials, Explosives, Fuels and Lubricants
Großes Cent, D-53913 Swisttal (Germany)

Introduction

Currently the armed forces in Germany are developing a Mobile Mine Detecting and Clearing System with the abbreviation MMSR. Three different types of sensors (EMI, GPR and MSP – working in a scanning mode) are applied in combination to find objects under, on or near the ground, the shapes of which correspond with that of Anti-Personnel Landmines (APL) or Anti-Tank Mines (ATM). None of these sensor technologies are able to identify the bulk material as explosives inside the object. As such, these techniques can not distinguish between a land mine and a variety of other man-made or natural objects or effects, often referred to as clutter. This results in an unacceptable number of false alarms, in many cases coupled with a dangerously low probability of detection. For that reason it is necessary to have a Fail-Safe-Sensor in order to prove the results of the mine detecting processes.

Within the framework of so-called humanitarian demining the requirements for all mines to be identified and cleared implies the necessity for all alarm signals to be properly investigated, which results in a dangerous and slow process. For military use a mine detecting device should work in real time to protect a convoy or other units. The only mobile sensor system which works as a biochemical sensor and correlates the odour recognition of explosive related chemicals (ERC) with the presence of buried mines is a well trained dog. However, canines employed for this mission have a relatively short attention time and they get tired very quickly. To solve that problem specific chemical sensors or arrays of sensors (electronic noses) should be evaluated and proved to be capable for the detection of land mines.

Vapor signature of buried landmines

To detect landmines with the help of chemical sensors it has to be assumed that mine related specific chemicals – i.e. explosive related chemicals - are present in the vapor phase above buried mines. A lot of well known parameters and physical/chemical processes such as composition and quantity of ERC, their partitioning (soil, water, air), various climatic events etc., affect the concentrations of the ERC on the soil solid phase and in the liquid phase. When the ERC have reached the soil surface area, the boundary conditions such as plant growth and transpiration, precipitation, atmospheric conditions (p, T, RH), thermal radiation and wind affect the chemical signature.[1]. All these land surface processes may play a significant role in the horizontal and vertical transportation of chemicals. Estimated boundary

[1] S.W. Webb, J.M. Phelan, „Effect of Diunal and Seasonal Weather Variations on the Chemical Signtures from Buried Landmines/UXO“, „Progress on determining the vapor signature of a buried landmine“, Proceedings of SPIE Vol.4038 (2000), S. 474..488

H. Schubert and A. Kuznetsov (eds.), Detection of Explosives and Landmines, 133–136.

layer concentrations at 22 °C using soil/air partition coefficients from moist soils (100% w/w) are 0,02 ppb for 2,4 DNT and 0,1 ppt for 2,4,6 TNT [2]. Moreover buried mines are unlikely to impart chemical vapor detectors with a volume of saturated air. At best, the fraction of molecules available to detectors will be reduced further by a few orders of magnitude. Estimates for this specific application are that useful detectors will need to be sensitive to 1 to 300 fg. An additional difficulty exists when the mine detecting process will be performed in a post blast environment. All these parameters have been discussed in the literature and indicate that the detection of explosives and explosive related chemicals within this particular application is still a big challenge for chemical vapor detectors. To evaluate the wide variety of detection methods and trained explosive detection canines, we have developed mobile test samples which can be used in different environments. Trace amounts of explosives and explosive related compounds should be detected with sensitive and selective techniques with short response times.

Test-samples for explosives and explosive related chemicals

The modern instrumental analytical chemistry is capable to identify lowest quantities of chemical substances. With the location of explosives as well as objects with explosives in different real cases like buildings, vehicles, persons, above soil (mines), nevertheless one is dependant on efficient mobile well proved sensor technique.

The school for training dogs of the armed forces in cooperation with the Bundeswehr Research Institute for Materials, Explosives, Fuels and Lubricants (WIWEB) has defined certain requirements that a test sample, especially that with small quantities of explosives (micro-mass-test sample / MPK), should keep.

The MPK was defined as follows:

1. Through transportation of test samples, also in bigger quantities (at least 100 pieces) there exist no dangers with view of explosion, flash point, and virulence

2. While handling the micro-mass-test samples the rules of the national Explosives Law will not apply.

4. Different explosives can be deposited safely on the surface.

5. Persons, who work with the test sample, do not come in contact with the explosives.

6. The sample-holder as well as the contaminated surface can become inert, i.e. without background-smells of explosives.

7. Using exact geometry as well as defined quantities of explosives on a well defined surface different source-strengths of evaporating ERC can be constructed.

---

[2] V. George, T.F. Jenkins, J. M. Phelan, D. C Leggett, J. Oxley, S. W. Webb, P.H. Miyares, J.H. cragin, J. Smith, T.E. Berry, „Progress on determining the vapor signature of a buried landmine", Proceedings of SPIE Vol.4038 (2000), S. 590..601

8. The test samples can be hidden at different places. Their loss can be tolerated.

The manufacture of the micro-mass-test sample (MPK) gave up using organic materials for its manufacture . To make the test sample inert, its individual components were heated up to 600 K over a time period of 1 to 2 hours. It had to be guaranteed that a contact with the explosive is not possible. In fig. 1 the test sample is shown.

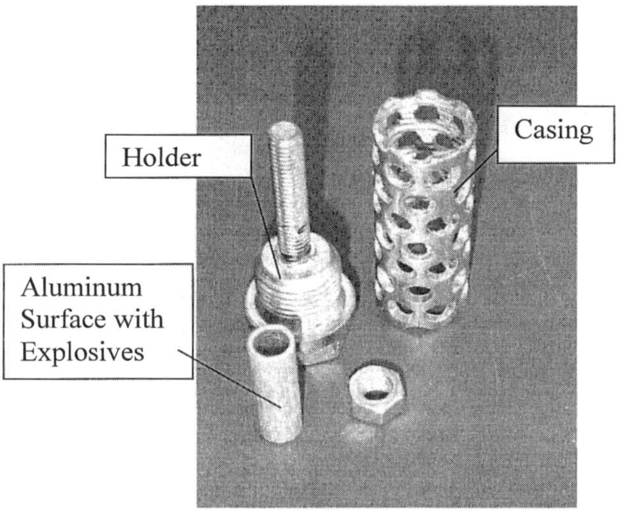

Figure 1: Micro-mass-test sample; the explosive was stored on the surface of the aluminum shells

To put the explosive on the surface of the metal-husk, the material must be shredded mechanically with a suitable procedure. The nature of the surface as well as the size-distribution of the explosive particles determine the loading characteristic. Investigations with HPLC show that the manufacturing variations are less than 10 wt. %. The deposit procedure can be performed without using additional chemicals at ambient temperature. At this time an exact procedure-description cannot be given for patent reasons.

The flux of TNT at 10 ° C is in the range of 100 pg/ min for the whole MPK when 8 mg of TNT were put on the aluminum surface (evaluation of DTA experiments). Using less quantities of explosives, the TNT flux is reduced to < 1pg/min which can be compared with measured values from test materials and landmines [2]. The TNT particles stick on the Al-surface. Pictures from the Scanning Electron Microscope (SEM) show the topography of the surface with TNT.

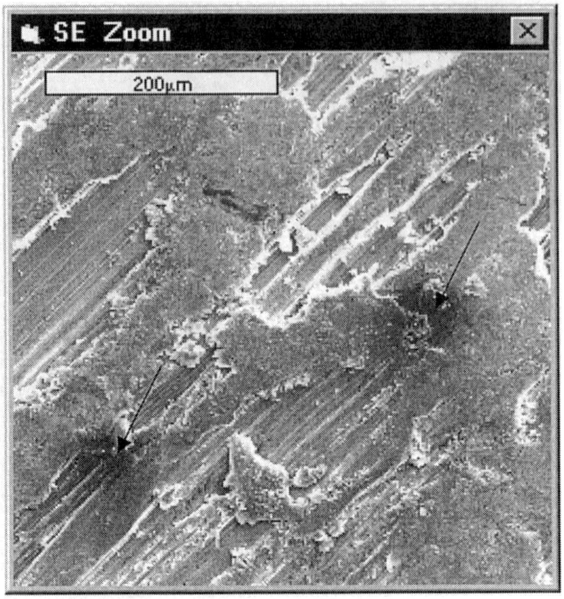

Figure 2: Topography contrast; chemicals with C-content were marked within the black regions

Conclusion

Within the framework of a round-robin test, in which different teams consisting of dog handlers with trained dogs and companies with sensor technology had participated, it was shown that the test samples, despite small quantities of explosives (micro-mass-test samples - MPK) are suitable for handling different defined sources of evaporating explosive related chemicals, to evaluate canine performance as well as different sensor systems. We are now investigating whether we can use these test samples for the training and the evaluation of the performance of dogs (quality assurance).

Acknowledgements

Financial support by the German Federal Office for Defense Technology and Procurement (BWB) is gratefully acknowledged.

EXPLOSIVES LOCALISATION AND PRE-IDENTIFICATION BASED ON UHF ELECTROMAGNETIC WAVES

V.P.AVERIANOV, A.S.VISHNEVESKI, I.B.VOROBIOV, M.D.ZUBKOV, A.V.KUZNETSOV
V.G. Khlopin Radium Institute
194021, 2nd Murinski pr., 28, St.-Petersburg, Russia.

1. Introduction

There are several ways in which ultra-high frequency (UHF) radio waves can be used for explosive' detection. Firstly, UHF radio waves have relatively high penetration ability, and can be used to quickly localize the "anomaly" in the soil. Secondly, they are sensitive to the electrical properties of the hidden object, which can be used to identify the nature of this object, i.e. to determine whether the object belongs to a certain class (for example: metal - not metal; dielectric with big dielectric permeability - dielectric with small dielectric permeability). The widely used pulsed radars are capable of fast localization of anomalies [Peters(1994), Moffatt(1976), Vitebsky(1997)] and give some hints to their classification on the basis of definition of geometrical parameters of these anomalies [Burgeous(1996), Carin(1998)], intensity of scattered radiation, and presence of natural resonances [Chen(1997)]. Use of short pulses with duration $\Delta t \sim 1$ ns with an equivalent band of frequencies $\Delta f \sim 1$ GHz has its advantages and disadvantages. The main advantages of the pulsed radars are:

- They work at frequencies for which absorption in soil is low, and thus the pulse can penetrate rather deep.
- They allow one to have powerful short pulses while keeping power consumption low.
- There are a lot of standard ready-for-use methods for data analysis.

The main disadvantages of using pulsed radiation are:

- Low spatial resolution $\Delta L \sim c\Delta t \sim 30$cm.
- Need to account for signal distortion due to dispersion of the propagation medium.
- Pulses do not have well-defined boundaries due to transition processes.
- Strict requirements to the emitter and receiver: no multiple reflections, decoupling of emitter and receiver, stability.
- High cost of generator of nanosecond pulses.

Systems with continuous radiation of variable frequencies can also be used for detecting anomalies [Robinson(1974)]. The main advantages of such systems compared to pulsed radars are:

- It is easy to provide a wide range of frequencies equivalent to the pulse length $\Delta t \sim 0.1$ ns, hence spatial resolution $\Delta L \sim c\Delta t \sim 3$ cm.

H. Schubert and A. Kuznetsov (eds.), Detection of Explosives and Landmines, 137–146.

- Possibility of off-line correction of amplitude- and frequency characteristics of emitter and receiver.
- Well-defined frequency boundaries of the signal, outside which it has zero energy.
- Less strict requirements to the quality of emitter and receiver, and to their decoupling.

The presented work was dedicated to the application of continuous UHF radiation in wide frequency range to localization and identification of explosives. The work was carried out in the framework of the ISTC Project #1050.

2. The scheme of the experimental setup.

In order to increase the spatial resolution we used high frequencies which usually are not applied to probing damp environments because of high absorption. In our measurements generators working in three frequency ranges were applied: 5 - 8 GHz; 8 - 12 GHz; 17 - 25 GHz. The power of radiation was ~ 1 mW, period of scanning each frequency range ~ 1 s. A dynamic range of sensitivity of the setup was ~ 35 dB, stability of signal at the level -30 dB was ~ 2%. Two independent horn antennae were used. The scheme of the experimental setup is shown at Figure 1.

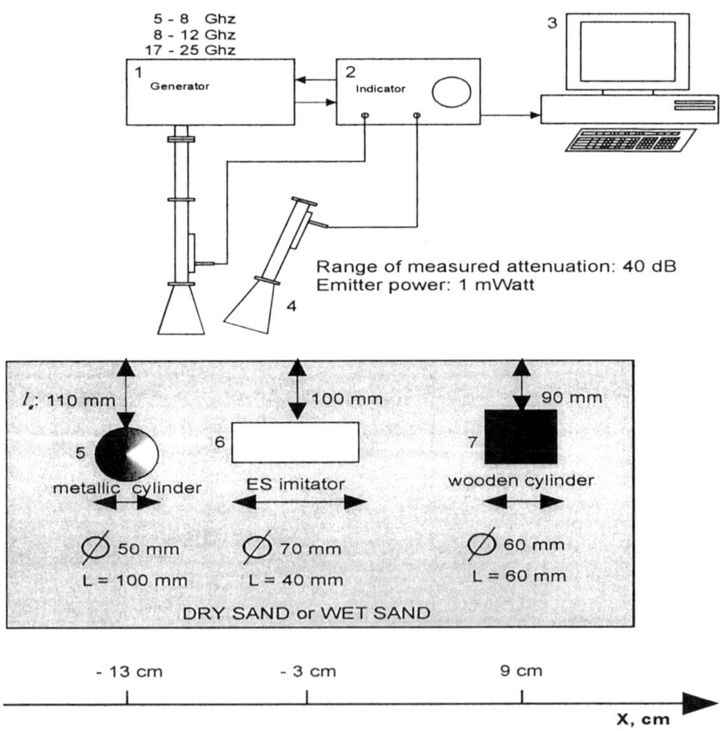

Figure 1. Drawing of the laboratory experimental setup.

The inspected objects were hidden inside the test container with ~40 kg of river sand. In one case humidity of sand was ~ 0.2%, in the other – 1% by weight. The simulator of explosive (TNT) was chosen on the basis of measurements of dielectric properties of real explosives; its dimensions were: diameter 70 mm, height 40 mm. Two other objects were used: wooden cylinder with humidity ~ 10% by weight having diameter 60 mm and height 60 mm, and metal cylinder with diameter 50 mm and height 100 mm.

The signal corresponding to the weakening of the reflected signal was detected and analyzed. The phase of the wave was not registered. The relative positioning of antennae provided their de-coupling at the level of -35 dB in case of dry sand, and - at the level of -25 dB in case of damp sand.

3. Description of the model used in the identification technique

The basic parameter describing the object is its dielectric permeability, which determines its reflection coefficient and electrical length. In a case of a dielectric with small losses, the electrical length $l_e = l_{ob} \cdot \sqrt{\varepsilon_{ob}'}$, where l_{ob} - length of the object, ε_{ob} - relative dielectric permeability. Hence, the problem of identification of the object can be formulated in the following way: on the basis of knowledge of the size and dielectric properties of the given object (e.g., an explosive) to find this object among others.

In order to proceed with identification, a model of the TNT imitator hidden in sand was created. It included the following parameters: dielectric properties of the environment (sand) and of the object, and position of the object with respect to the border between sand and air. Comparison of the result of calculations and experiment for the TNT imitator in sand is shown at Figure 2.

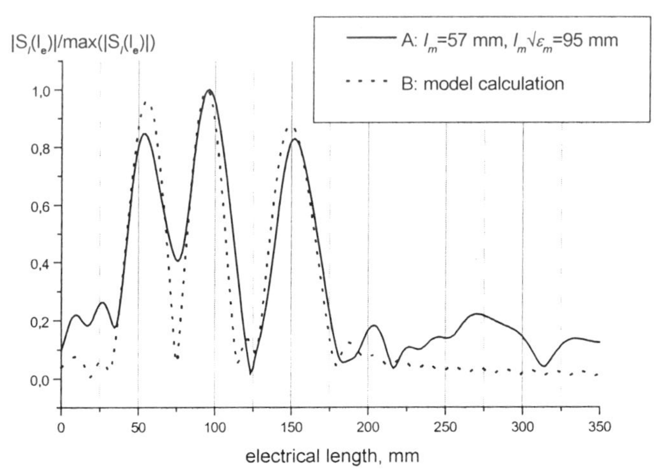

Figure 2. Comparison of calculated and experimental dependencies of module of normalized spectral density of the reflected signal on frequency. The hidden object was a TNT imitator placed at distance 55 mm from the border sand – air. The scanning frequency range was 17 - 25 GHz.

Fourier transformation of the measured reflected signal was used to reveal its periodical structure. Then, the Fourier-image of the modeled signal as a function of $l_m \cdot \sqrt{\varepsilon_m}$ was constructed for the known parameters of several hidden objects and the environment (sand). Correlation function of the measured and experimental Fourier-images at zero displacement was used as a criterion of identification.

4. Experimental results

Before any identification procedure using model results is used, stability of the electrical properties of the chosen TNT simulator in air must be examined, and its electrical length which is used in the model input must be measured. Figures 3, 4, and 5 show the dependence of the module of spectral density of the detected signal I(f) (Fourier spectrum) on the geometrical parameters of the experimental setup. The results indicate that there l_{ob} = *57.5* mm in the relatively wide range simulator displacements and rotation angles.

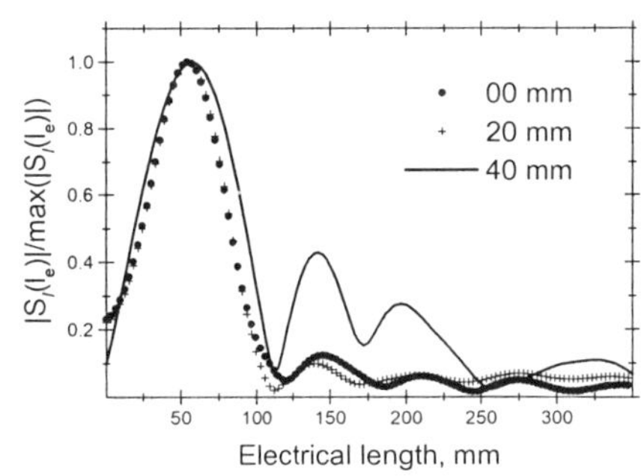

Figure 3. Stability of the Fourier spectrum towards the transversal displacement of the simulator relative to the symmetry axis of the antennae setup. The scanning frequency range is 8 – 12 GHz.

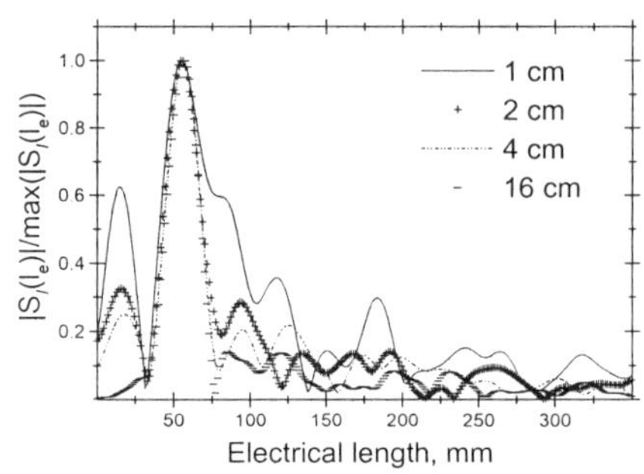

Figure 4. Stability of the Fourier spectrum towards changes of the distance from the simulator to antennae. The scanning frequency range is 17 – 25 GHz

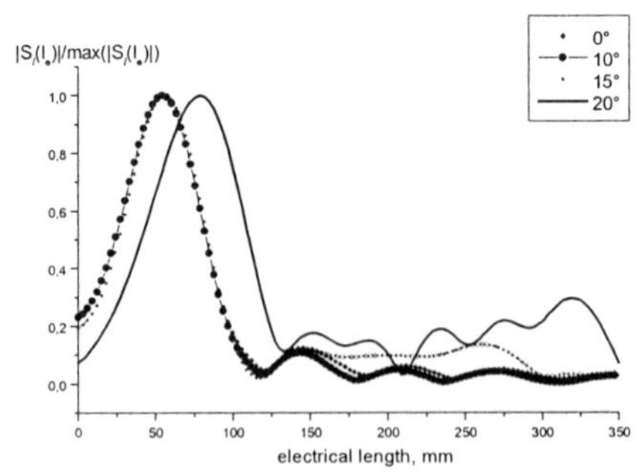

Figure 5. Stability of the Fourier spectrum towards rotation of the simulator around the symmetry axis of the antennae setup. The scanning frequency range is 8 – 12 GHz

The result of scanning of the test area (see Figure 1) along the X coordinate by the UFH radiowaves in the frequency range 17 – 25 GHz is shown on Figure 6.

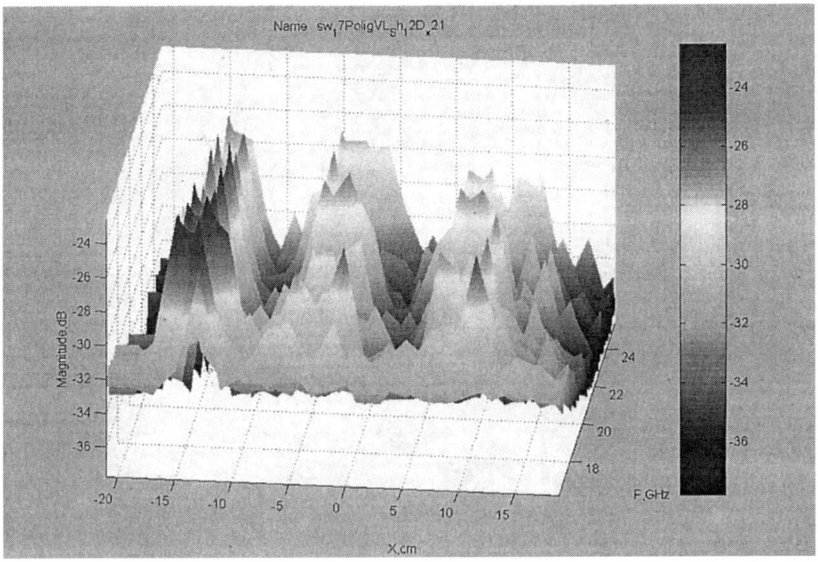

Figure 6. Visualization of the results of scanning of the test area in the frequency range 17 – 25 GHz.

The dependence of the integral weakening of the reflected signal on the X coordinate in shown at Figure 7 for two frequency ranges: 8 – 12 GHz and 17 – 25 GHz.

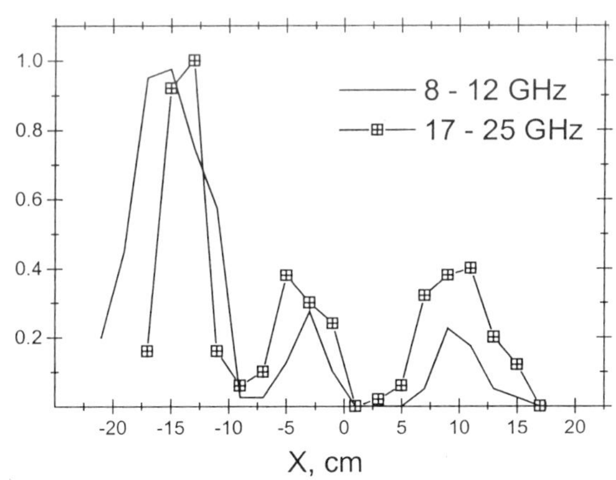

Figure 7. Result of integration of the normalized intensity of the reflected wave over the frequency range.

The most notable features of the Figure 7 are:

- position of hidden objects along the X axis is well reproduced;
- the intensity of the signal reflected from the metal cylinder with diameter 25 mm is several times higher than the intensity from two other objects;
- in the frequency range 17 - 25 GHz, for which the spatial resolution is higher, the peak corresponding to the metal cylinder becomes narrow (quasi-pointlike object),

while the width of the peaks corresponding to flat objects approaches their actual dimensions.

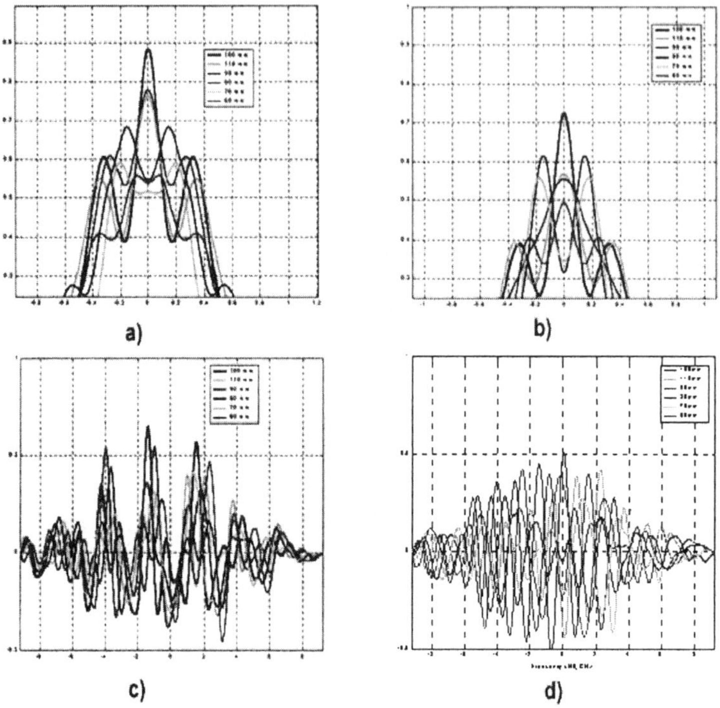

Figure 8. a) - correlation functions of the signal measured at coordinate X = -3 cm (TNT) and signal obtained by the model calculations for TNT at various $l_m \cdot \sqrt{\varepsilon_m}$; b) correlation functions of the signal measured at coordinate X = -3 cm (TNT) and signal obtained by the model calculations for metal at various $l_m \cdot \sqrt{\varepsilon_m}$; c) same, as a), but not for Fourier spectra but for the initial signal I(f); d) same, as b), but not for Fourier spectra but for the initial signal I(f).

Figure 8 shows correlation functions of the measured signal at coordinate X = -3 cm and calculated signals for different hidden objects. It is clear that correlation of the measured signal is the highest when the correct (TNT) hypothesis is used. Similar correlation functions for measured signals at coordinates X = -13 cm (metal) and X = 9 cm (wood) are shown at Figure 9.

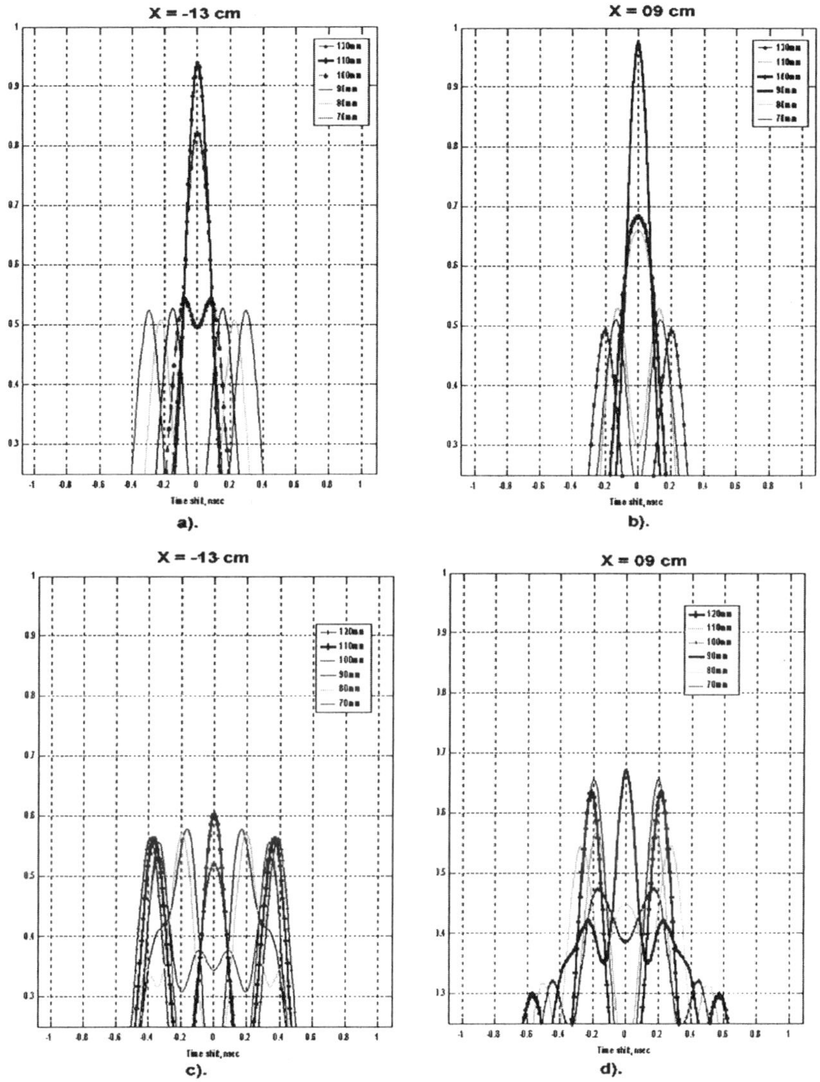

Figure 9. a) - correlation functions of the signal measured at coordinate X = -13 cm (metal) and signal obtained by the model calculations for metal at various $l_m \cdot \sqrt{\varepsilon_m}$; b) same for X = 9 cm (wood) and model signal for wood; c) correlation function for measured metal – modeled TNT combination; d) correlation function for measured wood – modeled TNT combination.

Figure 9 a), b) indicates that there are metal and wood reflecting surfaces at coordinates X = -13 cm and 09 cm and at depths l_e = 110 mm and 90 mm respectively. The hypothesis of TNT produces significantly lower level of correlation (Figure 9, c), d)).

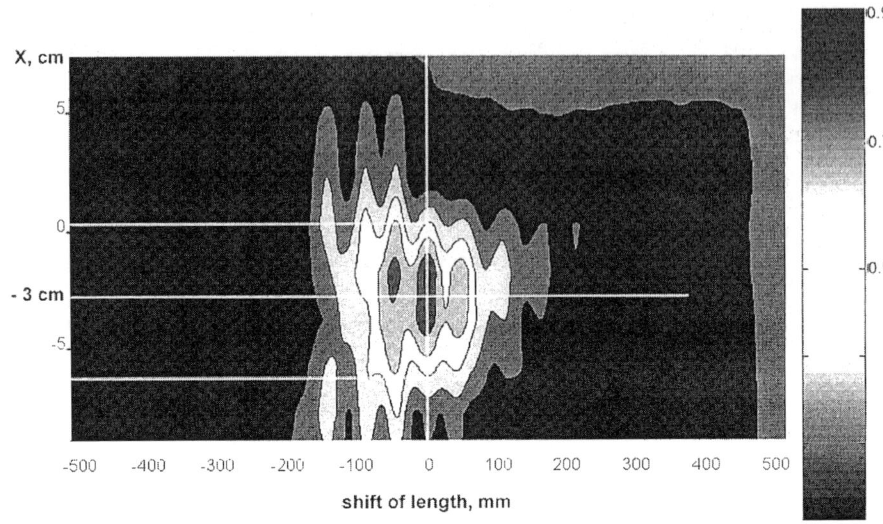

Figure 10. Results of the correlation analysis of the area between X = -9 cm and X = 5 cm.

In parallel with the correlation analysis, we have employed another technique for identification of hidden objects on the basis of weakening of the reflected radiowave – the analysis of common variations of spectra in space with Mahalonobis metrics (PCA/MD) [Cooley(1971), Gnanadesikan(1977) , Sharaf(1986)]. This method allows obtaining the probability that the unknown spectrum belongs to a specific group of spectra (the training set) corresponding to a particular hypothesis. The PCA Method compresses the spectral data and decomposes spectra from the training set into common variations characterized by an orthogonal set of eigenvectors. As a result of decomposition of the measured (unknown) spectrum into this orthogonal set of eigenvectors, one gets a small set of well-defined weights showing a degree of presence of given variations in the unknown spectrum. The method of Mahalonobis distances allows one to calculate the "distance" between the unknown spectrum and the average from the training set in units of standard deviations, obtained from the analysis of borders of the training group in the space of eigenvectors.

The main advantage of the PCA/MD method is that it takes into account both common variations of spectra at one frequency and correlations between points of a spectrum at various frequencies (i.e. accounts for covariations). In other words, all the information contained in the spectrum is taken into account. The given method allows to carry out reliable probability calculations, including multidimensional cases. By including into the analysis a limited number of the first orthogonal components it is possible to filter the statistical noise which is contained in less significant higher-order components. Results of numerical experiments have shown that for the given conditions the optimal number of spectra in a training set is from 6 to 10, and the optimal number of orthogonal components is from 3 to 5.

Figure 11 shows the the degree of a deviation of experimental spectra of various samples from the training set consisting of spectra constructed on the basis of the model (see section 3), in the units of standard deviations. It is clearly visible from the Figure that PCA/MD can reliably identify TNT imitator.

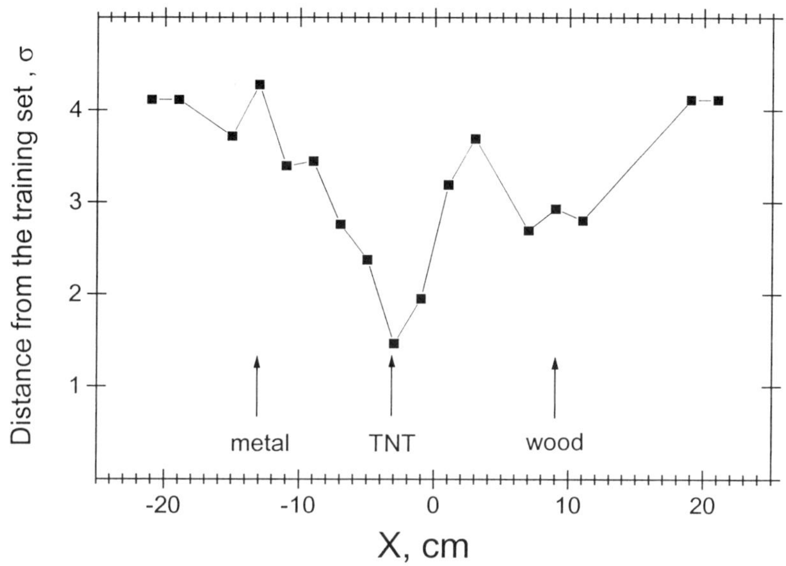

Figure 11. Result of identification of the TNT simulator by the PCA/MD technique.

5. Conclusion

It was experimentally shown that metals and dielectrics, including explosives, can be efficiently localized with continuous ultra-high frequency radiation. This technique allows one to carry out pre-identification of the hidden object. A procedure was developed to identify dielectric objects, including explosives in non-metallic cases. This procedure was applied to cases of high attenuation and dispersion (wet sand).

The method will be further developed to achieve good results for media with higher humidity. This will be achieved by introducing a reference channel, increasing the sensitivity of the receiver, optimizing the antennae system, introducing 2-dimensional correlation and statistical analysis, and using a wider range of samples and ES imitators and more precise models.

References

Peters Jr. L., Daniels J. J., Young J. D., (1994), Ground penetrating radar as a subsurface environmental sensing tool. Proc. IEEE, vol. 82, pp.1802-1822

Moffatt D. L., Puskar R. J., (1976), A subsurface electromagnetic pulse radar. Geophis., vol.41, pp. 506-518.

Vitebsky S., Carin L., Ressler M. A., Le F. H., (1997) Ultra-wide-band, short-pulse ground-penetrating radar: Simulation and measurement. IEEE Trans. Geosci. Remote Sensing, vol. 35, pp.762-772.

Burgeous J. M., Smith G. S., (1996), A fully three-dimentional simulation of a ground-penetrating radar: FDTD theory compared with experiment. IEEE Trans. Geosci. Remote Sensing, vol. 34, pp.36-44.

Carin L., Kapoor R., Baum C. E., (1998), Polarimetric SAR imaging of buried landmines. IEEE Trans. Geosci. Remote Sensing, vol. 36, pp.1985-1988.

Chen C. C., Peters L., (1997), Buried unexploded ordnance identification via complex natural resonances. IEEE Trans. Antennas Propagat. vol.45, pp.1645-1654.

Robinson L. A., Weir W. W., (1974), Location and recognition of discontinuities in dielectric media using synthetic RF pulses. Proc. IEEE, vol. 62, pp. 36-44.

Cooley W. and Lohnes P., (1971), Multivariate Data Analysis, Wiley, NY

Gnanadesikan R., (1977), Methods for Statistical Data Analysis of Multivariate Observations, Wiley, NY

Sharaf M., Illman D., Kowalski B., (1986), Chemometrics, Wiley, NY

A NEUTRON-GAMMA METHOD AND APPARATUS FOR DETECTION AND IDENTIFICATION OF HIDDEN OBJECTS IN BRICK (CONCRETE) WALLS

MIKHAIL N. CHUBAROV
V.G.Khlopin Radium Institute
28, 2nd Murinsky Av St-Petersburg, Russia.

1. Introduction

When considering problems of fighting terrorism, it may be of interest to imagine a situation when explosive materials or mines are hidden inside walls of buildings, pillars, piers and other barrier-type constructions made of materials like bricks and concrete. When the barrier can be accessed only from one side, detection of objects hidden inside it becomes similar to detection of land mines. Various techniques have been proposed to detect and identify explosives buried in soil. Most of them use penetrating radiation having qualitatively different characteristics of interaction with components of explosives and surrounding materials. The source of primary radiation and the detector are placed on the same side of the barrier and scattered primary or secondary radiation is registered. The most frequently used sources of primary radiation are sources of neutron, γ-rays and electromagnetic waves. An apparatus of this type makes it possible to detect foreign objects in soil and some of them (mainly those using neutron sources) are able to identify hidden materials by their elemental composition.

The common drawback of methods and apparatus of this type is rather small depth of detection usually limited to 15-20 cm, which is due to absorption of both primary and secondary radiation, and strong dependence upon the distance to the hidden object.

With barrier type objects of inspection it becomes possible to access them from both sides and source of radiation may be placed opposite to the detector with the inspected barrier between them. In this case the detector may detect radiation transmitted through the barrier without any interaction or radiation scattered at small angles with small energy losses. Measurements performed in such geometry may increase the depth of the detection and in certain situations make it possible to identify a number of materials including explosives.

To achieve not only detection but also identification of hidden objects one should use two or more kinds of radiation having different characteristics of interaction with hidden and surrounding materials. Neutron- and γ-radiation are quite suitable for this purpose and can complement each other. Fast neutrons from isotopic sources interact with atomic nuclei and are effectively scattered and slowed down by nuclei of light elements. Gamma radiation of isotopic sources interacts with atomic electrons and is strongly absorbed and scattered in heavy materials. As any neutron source emits gamma rays it seems to be promising to use both kinds of radiation from a single source, for example Cf-252. It would also be efficient to use in a similar way a single detector for both neutrons and gamma photons capable of counting them separately and simultaneously. Such detectors are widely used in nuclear physics. Pulses from

H. Schubert and A. Kuznetsov (eds.), Detection of Explosives and Landmines, 147–153.

scintillation detectors with certain types of organic or liquid scintillators have different amplitudes and shapes for electrons produced by γ-rays and heavy particles like recoil protons produced by neutrons scattered by hydrogen atoms of the scintillator. With an appropriate simple enough pulse shape discriminating electronic circuits it is possible to distinguish between neutrons an γ-rays ant count them separately, each in it's own channel.

An apparatus using a single neutron-gamma source and a single detector for both kinds of radiation is inexpensive and simple in use. Below we present some results of our work devoted to the development of such apparatus and a simple method of detection and identification of objects located within a uniform barrier-type constructions in situations when they can be accessed from both sides.

2. Experimental setup

Block diagram of the experiment is shown on Figure 1. Shielding container 20 cm in diameter and 25 cm in height with Cf-252 neutron source yielding 2×10^5 n/s ($\sim2\times10^6$ γ/s) was placed close to an 80×80×60cm brickwork. During the measurements the source was moved to its surface. The thickness of the brickwork varied from 0 to 60cm. The scintillation detector with organic scintillator (stilbene) was positioned opposite the source of radiation on the other side of the brickwork. Samples were placed in cavities inside the brickwork at different distances from the source, which were usually multiples of a single brick dimensions.

Pulses from the detector following amplification and shaping were fed to the pulse shape discriminator and then to neutron and gamma counters. Energy thresholds for counting gamma photons and neutrons were 0.5 MeV and 1.5 MeV correspondingly. Neutron-to-gamma ratio and plots of net neutron count versus net gamma count were used for detection and identification of hidden materials.

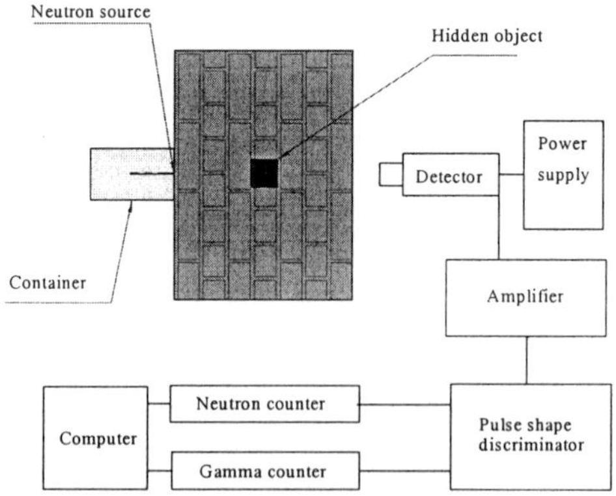

Figure 1. Block diagram of the experimental installation.

3. Results

To evaluate and demonstrate the potential of the method, absorption curves for neutrons and gamma rays of Cf-252 in different materials were measured. Layers of these materials were placed between the source of radiation and the detector. During these measurements source-to-detector distance was kept constant.

Absorption curves obtained for neutrons and gamma rays of Cf-252 in bricks, polyethylene and explosive (imitation of TNT) are shown in Figure 2.

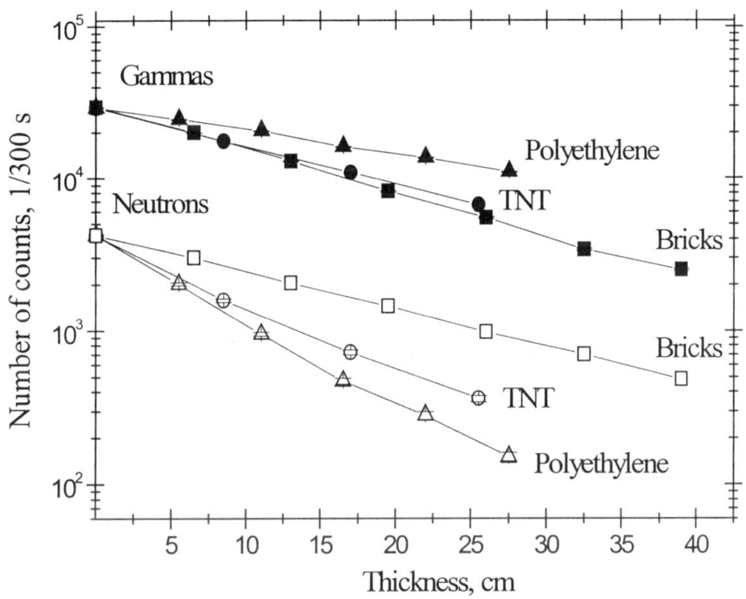

Figure 2. Absorption curves obtained for neutrons and gamma rays of Cf-252 in bricks, polyethylene and explosive (TNT) imitator.

All the curves are nearly exponents with absorption coefficients different for different materials. It should be noted that for attenuation of neutron and gamma radiation in bricks these coefficients (slopes of the lines) have nearly the same values. Next, for bricks and TNT gamma rays absorption coefficients differ only slightly. It means that when only gamma radiation is used for the interrogation the sensitivity of explosives detection in brickwork would be low. TNT and polyethylene attenuate neutrons more readily than bricks, which make them more promising for detection of these materials in brickwork. Difference between TNT and polyethylene (similar by its absorption properties to water) is more pronounced for gamma rays. In any case data on both kinds of radiation complement each other and indicate the possibility of detection of TNT and water in brickwork.

In real situations the source of radiation and the detector are placed right on the surfaces of opposite sides of the barrier and the intensity of radiation will vary not only because of the absorption but also with the source-to-detector distance depending on the thickness of the barrier.

Net neutron and gamma count rates versus brick barrier thickness measured in such geometry are plotted in Figure 3.

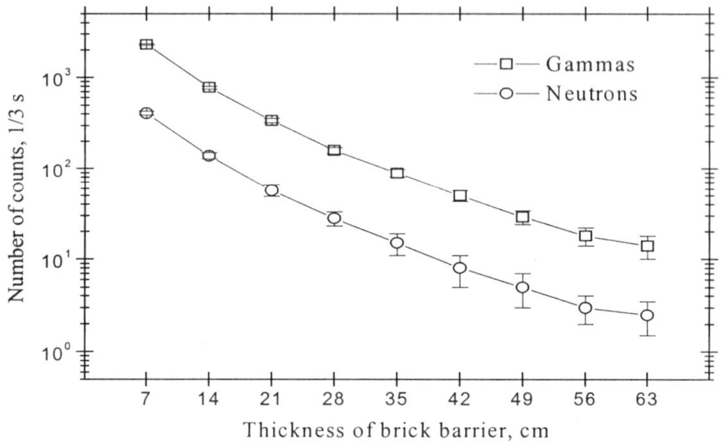

Figure 3. Net neutron and gamma count rates versus brick barrier thickness.

The relationship is nonlinear, but neutron-to-gamma intensity ratio remains essentially constant for radiation transmitted through barriers and does not depend upon their thickness. Therefore neutron-to-gamma plot yields a straight line. This plot is shown on Figure 4 for barrier thickness from 10cm to 60cm. This feature is remarkable as it broadens the possibilities of detection and identification of foreign materials in brickwork.

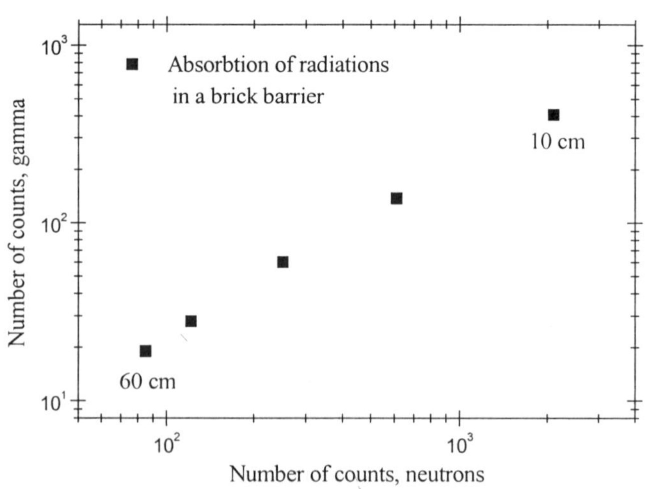

Figure 4. Neutron-to-gamma yields for different thickness of the brickwork.

It should be noted that data at Figures 3 and 4 are plotted after removing the natural background. This procedure is essential for low count rates when working with thick barriers. The background count rate may depend upon the radioactive contamination of the building materials and depends upon the thickness of the barrier. One more essential feature is that after the proper subtraction of the background the straight line crosses coordinate axes at the origin.

To evaluate characteristics of detection and identification of different materials hidden within the brick barrier, samples of polyethylene, steel, lead, and TNT were placed in cavities at different distances from the source. The samples were 6.5×6.5 cm plates. Their thickness could be varied from 0 to 6.5 cm.

In Figure 5 the number of neutrons is plotted against the number of γ-rays for steel plates of different thickness placed at different depths in 60 cm brickwork.

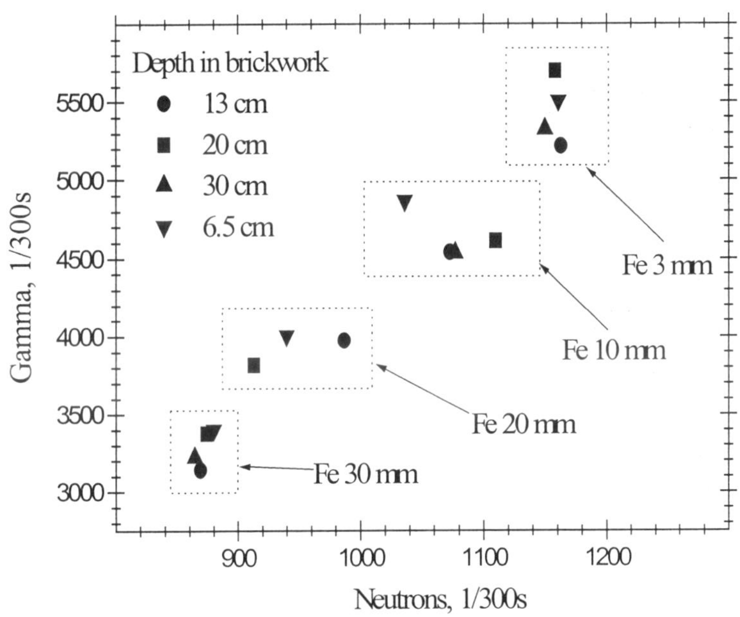

Figure 5. Number of neutrons versus number of γ-rays for steel plates of different thickness placed at different depths in 60 cm brickwork.

The most remarkable feature of the plot is that the number of neutrons and gamma quanta transmitted through the brick barrier strongly depends upon the thickness of the steel plates and far less upon their position inside the brickwork. Points corresponding to the same thickness of steel plates at different positions in the brickwork are grouped in common regions of the plot. Similar results were obtained for other materials. This feature facilitates their identification.

Plot of neutron against gamma counts for steel and other materials of different thickness positioned at different depths in 60 cm brickwork are shown in Figure 6.

Points on the plot correspond to the thickness of polyethylene varying from 3 to 60 mm, steel from 3 to 30 mm, lead from 3 to 24 mm, and TNT 30 and 60 mm.

One of the most important elements on Figure 6 is the straight line corresponding to the solid brick barrier with empty cavities of different height. This line is similar to that shown in Figure 4 for barriers of different thickness. It may be drawn through the point corresponding to the solid barrier and the origin taken at zero point. This line divides the plot into two regions. The upper one corresponds to light materials hidden in the brickwork and the lower part – to metals and other heavy materials. Neutron to gamma ratio may be used for identification of materials and absolute number of counts may be useful for evaluation of their mass (thickness). Sensitivity of detection and reliability of identification is determined primarily by the amount of hidden material.

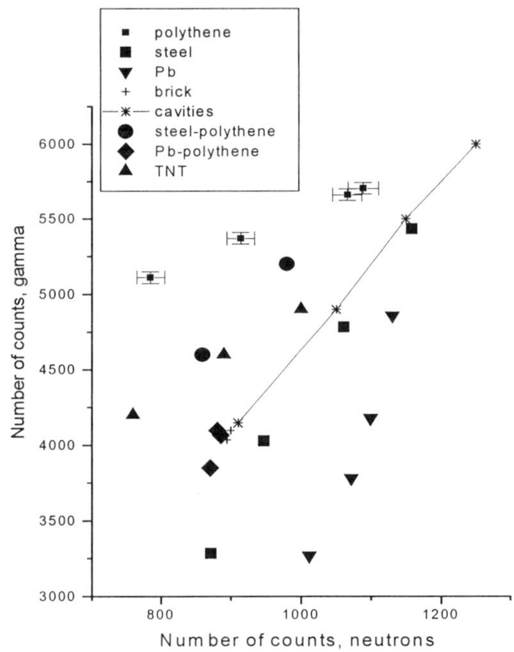

Figure 6. Neutron-to-gamma counts for steel and other materials of different thickness positioned at different depths in 60 cm brickwork

Neutron to gamma ratios for mixtures of light and heavy materials will differ from those for pure ones and points on the graph corresponding to them will shift relative to the dividing straight line. In Figure 7 two points denoted by filled circles correspond to 3 mm and 10 mm steel plates and 30 mm polyethylene plate placed together in a cavity in the brickwork at 30 cm from the source. It is clear that steel and polyethylene or TNT at similar proportions would be taken for some light material. Moreover, certain combinations of materials hidden within brickwork will not be detected, as for example 10 mm plate of lead in combination with 30 mm plate of polyethylene (points denoted by triangles in Figure 2). In such cases it may be necessary to calibrate the apparatus with the expected mixtures and use for their identification not only neutron to gamma ratio but also some more sophisticated mathematics like pattern recognition technique. In any case, such points should be treated as suspicious and one should try to determine their shape or use some other technique to check the result.

Results of detection of 0.5 kg of TNT in 20 – 60 cm brick wall are presented in Figure 7.

When the thickness of the barrier increases, the contrast (ratio of difference in counts with and without TNT to total count) softens slower for neutrons (from 20% to 12%) than for gamma rays (from 20% to 6%). For reliable detection and identification of 0.5 kg TNT in 60 cm brickwork, the statistical accuracy should be of the order of 1%. It means that with 2×10^5 n/s Cf-252 neutron source it would take about 300 s per one measurement for 60 cm brick wall. For 2×10^6 n/s neutron source this time will be about 30 s and 1% accuracy may be achieved for 1m barrier in 300 s.

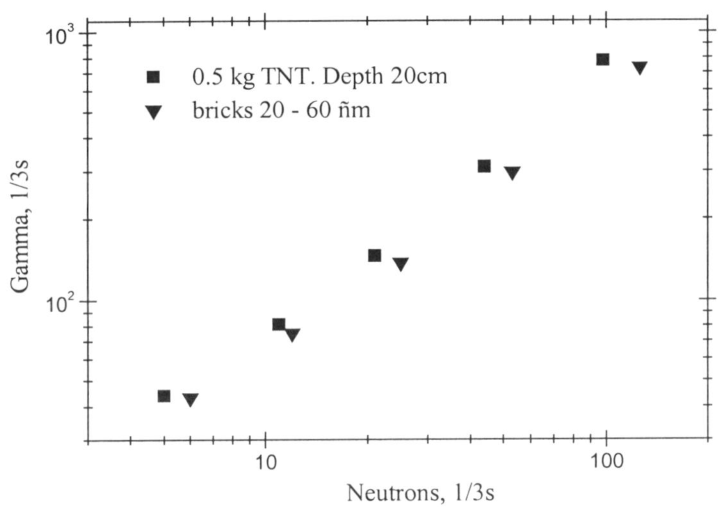

Figure 7. Results of detection of 0.5 kg of TNT in 20 – 60 cm brick wall.

Similar measurements have been made for barriers of concrete and the results are qualitatively similar to that for bricks.

4. Conclusions.

Method and apparatus using single neutron source and single detector for neutron and gamma radiation is proposed for detection and identification of foreign materials (including explosives) in barrier-type constructions made of bricks, concrete and other building materials. Sensitivity of explosives detection in 60 cm brick wall is 0.5 kg of TNT in 35s with 2×10^6 n/s Cf-252 neutron-gamma source and 1 kg of TNT in 1m wall in 360s with the same source of radiation. Apparatus of this type is cheap, portable and simple in use. It may be more effective in combination with other devices based on different principles of detection of explosives.

Practical Issues in Manual Demining: Implications for New Detection Technologies

James Trevelyan
Department of Mechanical and Materials Engineering
The University of Western Australia
Nedlands 6907
E-mail: jamest@mech.uwa.edu.au
Web: http://www.mech.uwa.edu.au/jpt/

Summary

New landmine and explosive detection technologies have been the goal of many research groups since the mid 1990's. Apart from obvious military advantages, the motivation has been a desire to speed up landmine clearance rates significantly. Current landmine clearance relies mainly on metal detectors and the false alarm rates are very high. It is widely believed, correctly, that this creates time-consuming work for deminers. However, analysis of actual demining rates shows that the clearance rates are faster than many people might expect, and that factors other than detection of targets can dominate performance. Vegetation clearance can take as much as 75% of the time needed by deminers.

This paper presents a simple model of minefield clearance performance which seems to reasonably represent recent field data. Afghanistan mine clearance statistics suggest that manual deminers clear minefields at a typical rate of 150 sq m per hour (for one breeching party of two men working alternately), plus an extra 2.5 minutes, on average, for every metal target encountered in the minefield.

Using the model, we present some calculations on the likely cost savings which could result from advanced technology hand-held detectors. A detector that eliminates 99% of false alarms could improve clearance rates by about 60% in typical vegetation conditions of Cambodia and about 3 times in typical Afghanistan conditions where vegetation is usually not a major problem. These improvements are not as much as some might expect.

The main implication of this work is that the affordable price of new generation detectors might be considerably lower than expected.

Introduction

How much time does it take a deminer, on average, to investigate and remove a typical metal fragment once it has been detected using a metal detector?

The sequence of manual demining operations needs to remembered to place this part of the operation in context. Details vary, but the procedure is more or less the same in most demining operations.

H. Schubert and A. Kuznetsov (eds.), Detection of Explosives and Landmines, 155–164.

1. The team lay out the minefield perimeter, mark using painted wooden posts and/or stones and rope or tape.

2. The deminers prepare breeching lane entry points, marking the starting position with a (typically) 1 metre long wooden cleared-position-marker rod.

3. The deminer checks for trip wires in next 0.5 metres of his lane, both visually and using a thin rod to lift tripwires lying on the ground.

4. He cuts and removes vegetation in next 0.6 metres of his lane, checking for trip wires not removed already.

5. He lays a wooden marker rod 0.5 metres from current cleared position to mark the area to search with the detector.

6. He searches the area between the two marker rods 0.5 metres apart for targets using his detector.

7. As he finds each signal, he lays the detector down on the ground, places a marker on the location, and checks the detector signal again to ensure the marker is in the right place. He determines where the signal is weakest between the marker and the current cleared position. He marks this point as well.

8. Using a prodder he carefully excavates the surface layer first, working forward from a point about 15 - 20 cm from the main marker. If he finds no metal target he repeats this, excavating deeper in layers until the target is found or the depth reaches a preset limit. In some operations excavations continue to a depth of 30 cm.

9. If (as is usually the case) he finds a metal fragment he removes it and checks the position again with a detector before moving on.

10. In some operations in soft ground, he carefully searches the main marked by probing with a long steel rod. However one cannot be certain of clearance to a depth more than 10 cm in ideal sandy conditions.

11. If a mine is discovered the mine is marked and the lane is closed off until a demolition charge can be used to destroy the mine *in-situ*.

12. Once the 0.5 metre deep area between the current cleared position and the marker rod has been shown to be metal free, the deminer moves the lane markings forward to the marker rod location. He also moves the current cleared position marker rod. Then the deminer returns to step 3 unless the minefield has been completely cleared.

13. After every few metres of lane has been cleared, in many demining operations, the deminers re-check the most recently cleared section of the lane with detectors, marking and removing targets using steps 5 - 9. Clearly this procedure can only be used when metal targets are removed. If the probing procedure is used and targets are left *in-situ* the targets must be marked until this re-checking is completed.

We investigated minefield clearance performance using data collected by the Mine Action Program for Afghanistan (MAPA). This data is collected on a routine basis for every minefield. We have used the clearance data which reports the time required to clear the minefield (hours), and the number of anti-personnel mines, anti-vehicle mines, unexploded ordnance and fragments found in the minefield. The data also records the land use and surface condition of the minefield and the method of clearance (manual, dogs, mechanical, or a combination).

In 1998 we derived a nonlinear statistical model of the data which predicts clearance rate as a function of minefield area, fragment density, and device density (anti-personnel mines + anti-vehicle mines + unexploded ordnance items). The original data for approximately 2000 minefields cleared between 1993 and 1997 were classified as follows:

Manually cleared minefields in agricultural and grazing land with soft ground
Manually cleared minefields in agricultural and grazing land with hard ground
Manually cleared residential minefields

Data for the following classes were excluded:

Manually cleared irrigation canal minefields
Manually cleared roads
All minefields cleared by dogs and mechanical support teams

A minority of minefields were cleared by more than 1 team. The data for these minefields were excluded because we could not ascertain whether the hours reported represented elapsed time or the total team-hours expended.

We were also advised that the data contained many errors. After we analysed this data we took a closer look at data representing minefield boundaries. We found many data entry errors showing that it had not been possible to eliminate consistency errors from the data base. (Much of the data have since been corrected.)

Trevelyan (1998) reports the details of the nonlinear statistical model. Different models were constructed for each class of minefield listed above, and different sets of coefficients were obtained from the data. Statistical models seldom reveal much information about causes and effects -- this is a common limitation. The different sets of coefficients for the different classes of minefield did not tell us much about which factors might be significant. It was not easy to see whether there was a significant difference, for example, between soft and hard ground conditions. However, we could conclude that the number of explosive items found in the minefield had a negligible effect on clearance rate.

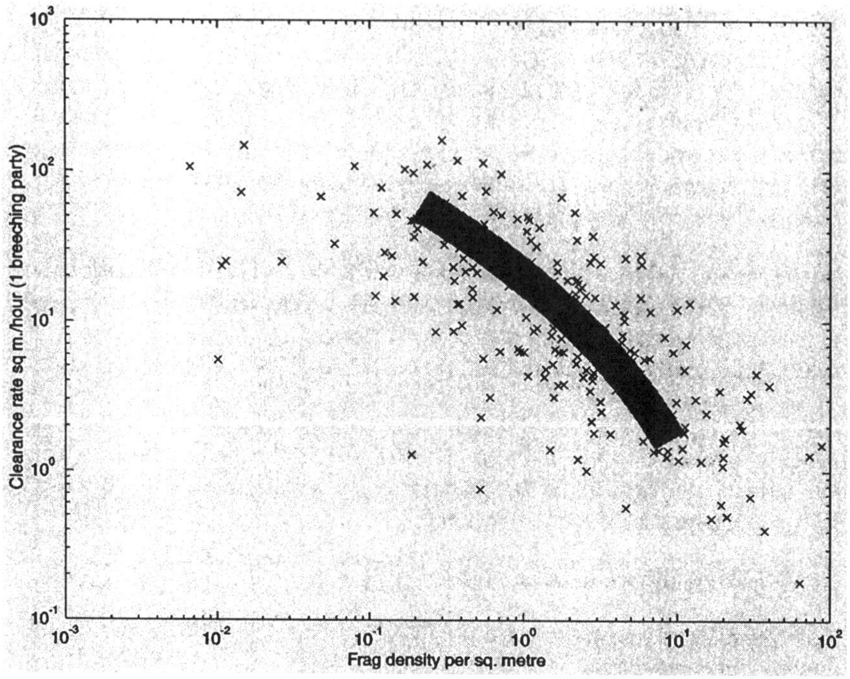

Figure 1: Logarithmic plot of clearance rate and fragment densities for manually cleared residential minefields in Afghanistan. The thick line represents a "line of best fit" by eye.

Recently we took a different approach to modelling and obtained some very interesting and surprising results.

Figure 1 shows a logarithmic plot of clearance rate (vertical axis) and fragment density (horizontal axis). The clearance rate has been calculated for a single breeching party of two men working alternately. Similar plots were produced for each class of minefield listed above. A curve of "best fit" was drawn by eye for each plot.

Then we combined all the curves on a single plot and found, as figure 2 shows, to our surprise, that all the curves for manual clearance nearly overlapped each other. In other words, a single curve may satisfactorily represent all the data for manual clearance. From this we could conclude that there is little significant difference between the different classes of minefields.

The major difference between different classes of minefields is simply the range of fragment densities encountered. Most agricultural and grazing land has between 1 fragment per sq m and 1 fragment per 100 sq m. Most residential areas have between 1 fragment per 10 sq m and 10 fragments per sq m.

We constructed a simple "heuristic" model with two parameters to explain the clearance rate. The first parameter is a constant clearance rate in the absence of fragments. This rate is determined by the speed at which deminers can search the ground using metal detectors, mark the lane boundaries, move the markings, check their work, record their work, and attend to equipment failures. (Steps 1 - 6 and 12 - 13 of the demining procedure) The second parameter is an average time which is needed to locate and investigate each metal target (steps 7 - 9, and 10 and 11 when required).

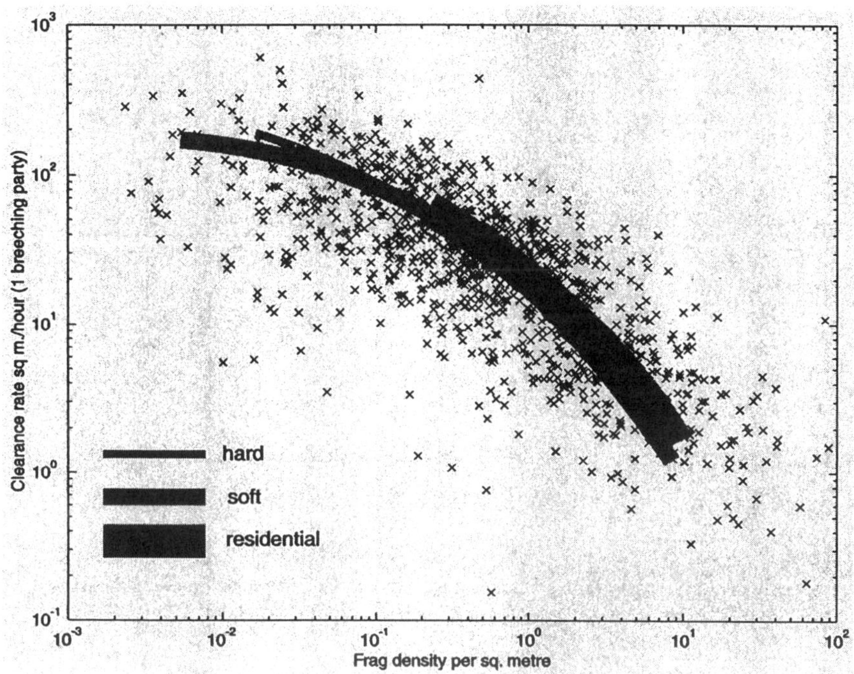

Figure 2: Logarithmic plot of clearance rate and fragment densities for 3 classes of manually cleared minefields, showing lines of best fit determined "by eye".

Figure 3 shows the manual clearance data overlaid by a curve representing a clearance rate of 150 sq m per hour (in the absence of fragments) and 2.5 minutes per fragment. Surprisingly this simple heuristic model fits the data almost as well as the nonlinear statistical model.

Field deminers are surprised by this result. However the result applies to a breeching party of two deminers, though only one is working at any one moment. On a per man basis, the zero fragment rate is therefore 75 sq metres per hour. Of course, in reality, fragments are always found in minefields so the actual clearance rates mostly much less as the graph above shows. Extremely fast rates are most likely explained by recording errors.

Recently we have analyzed clearance data recorded between 1997 and 2001, confirming these estimates.

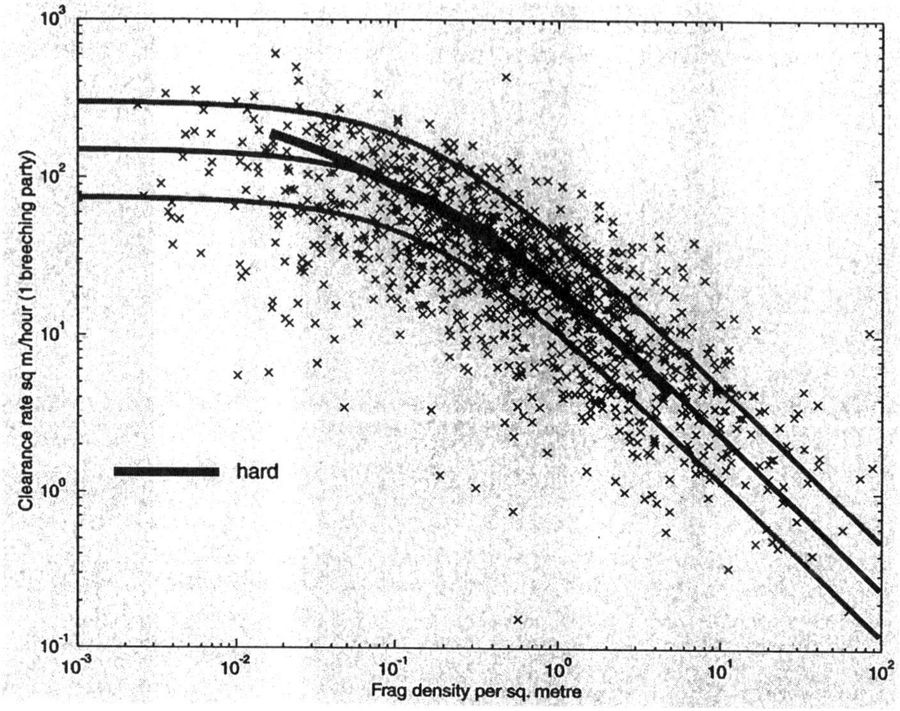

Figure 3: Logarithmic plot of clearance rate and fragment densities for 3 classes of manually cleared minefields, showing curves from a simple two parameter model. The lighter lines above and below the model characteristic represent twice the model rate and half the model rate respectively.

Further Experiments

At the time of writing we are conducting simulated demining experiments to test these results. Standard 25 mm square by 2mm thick steel targets are placed in a test field at random positions, at a given target density. The time required for metal detector detection, location, marking and excavation are being measured, together with the size of the excavation required.

One significant result is excavation time. For shallow (non visible) targets this is around 45 seconds using the standard operating procedure. The average detection and marking time is around 25 – 35 seconds. These results support the validity of the 2.5 minute average fragment recovery time obtained from the clearance data analysis.

Implications for Economics of Advanced Technology Detectors

These results suggest that the improvements which might be gained from advanced technology detectors are not as significant as one might expect. We would hope that an advanced technology detector would eliminate most of the false alarms by discriminating between fragment targets and mine-like targets.

Two frequently cited replacements for metal detectors are:

a) Combined metal detector and ground penetrating radar (GPR)

b) Combined metal detector and nuclear quadrapole resonance detector (NQR)

Both of these detectors require accurate target location using the metal detector channel. The GPR detector requires accurate positioning relative to the target and the ground for reliable discrimination between mines and metal fragments. The NQR detector requires high electric power levels that cannot be used continuously. Depending on the particular technology used, the accuracy of detector location required for the subsequent step (i.e. discriminating between metal fragments and explosive or mine-like targets) may vary. Here it is important to make two important qualifying remarks.

First, metal detectors do not always accurately determine the location of metal fragments. Very weak targets (typical of minimum metal mines) may be more easily detected at the edges of the coil rather than the centre. Also, when more than one fragment is close enough to be detected at the same position, the location errors are often large.

Second, there are many large metal fragments present in minefields, particularly in areas formerly occupied by people or battlefields. So far I have not seen data on the extent to which such fragments will disrupt the operations of GPR or NQR detectors. One can reasonably expect that a certain proportion of these larger fragments will have to be checked manually.

Let us assume then that the GPR and NQR detectors will still require part of step 7 in the demining sequence, even for targets classified as "safe". We do not yet know what proportion of metal fragments can safely be classified this way.

However we can suggest that these detectors will need a typical "discrimination" time for all targets, and an additional manual excavation time for targets classified as "unsafe".

Modelling Deminer Clearance Performance

We model the time required for one deminer (working full-time) to clear 1 square metre as:

$$1/v + 1/r + t^*f/60 \text{ (hours)}$$

Where: v is the rate of vegetation clearance (sq m/hour),
f is the fragment density (per sq m),
r is the rate of clearance (sq m/hour) in the absence of any targets, and,
t_f is the average time needed to locate, mark, investigate and clear one target (minutes).

The clearance rate is then given by $1/(1/v + 1/r + t_f *f/60)$ sq m / hour

In Afghanistan, deminers work in pairs: one rests while the other works and they alternate every 20 minutes or so. Therefore, these rates apply to one pair of deminers (a breeching party). Each manual demining team has 12 breeching parties.

In Afghanistan, vegetation is not a major problem in most minefields. The data suggests that the combined rate for vegetation clearance and area clearance in the absence of fragments is about 150 sq m/hour.

Given this model, we can make some predictions of the likely improvements obtainable with an advanced technology detector which can distinguish fragments from mines. If the proportion of fragments classified by this detector as "not mines" is x, and the time for the deminer to classify a fragment target as "not a mine" is t_1 then the improved clearance rate is given by:

$$1/(1/v + 1/r + ((1-x)\ t_f + x\ t_1)*f/60) \text{ sq m / hour}$$

We can then calculate the improvement possible for different values of x and f. This has been done in two tables below with the following parameters:

> r is the rate of clearance (sq m/hour) in the absence of any metal targets = 150 sq m/hr
> t_f is the average time needed to locate, mark, investigate and clear one fragment = 2.5 minutes
> t_1 is the average time needed to classify a target as a fragment = 0.25 minutes

Table 1 (next page) shows that the performance gains are greatest where the fragment density is greatest. However, the average fragment density for most minefields in Afghanistan is around 0.5 - 1 per square metre, and therefore the performance gain is likely to be about 2 - 4 times on average. Table 2 shows that the achievable gains for an advanced technology hand-held detector are quite modest if the deminers are slowed by having to clear vegetation. A gain of about 60 - 70% on present rates is predicted.

This in turn allows us to estimate the likely improvement in cost effectiveness.

A typical deminer in Cambodia and Afghanistan costs about US$10,000 per year to support in the field. This is calculated by dividing the overall cost of a demining program by the number of deminers in the field, which is a more accurate indication than the deminers' salary cost. The salary cost is a small component of this (approx $1800 per year). In Bosnia and Croatia, deminer costs are significantly greater, and costs up to US$40,000 have been given by some sources.

In a country with minimal vegetation problems, the performance gain and cost savings from using an advanced technology detector are significant. Given the costs mentioned above, savings of between US$20,000 and US$30,000 per deminer might be achievable. In a country with significant vegetation like Cambodia, but with a comparable cost structure, the savings could be as little as \$5,000 - \$8,000 a year. In a high wage cost environment, the savings are clearly greater.

Table 1: Improvement in demining rates for different factors x (proportion of fragments classified by detector as false alarms) and f (fragment density per square metre). This is based on clearance data from Afghanistan with negligible vegetation clearance time, and assumes that the deminer needs 15 seconds work to classify a target as "not a mine" using an advanced detector.

x\\f	0.1	1	10
0.5	1.21	1.63	1.80
0.9	1.45	3.31	4.93
0.99	1.52	4.31	8.13

Table 2: This table is based on the same parameters as above, but with a significant vegetation clearance times as in countries such as Bosnia and Cambodia (25 sq metres per hour per deminer).

	0.1	1	10
0.5	1.04	1.27	1.68
0.9	1.07	1.62	3.68
0.99	1.08	1.72	5.03

Mine detection dogs are increasingly being used in Afghanistan and elsewhere, though there are concerns about their performance in certain tasks. The overall clearance rates, using mine detection dogs, are much higher than in manual demining. The false alarm rates are much lower, and fewer fragments are removed by teams using these methods.

Once again, there is a significant implication for new detection technologies here. With lower false alarm rates, the relative advantage to be gained from new technology detectors is greatly reduced. In Afghanistan, there are plans to increase the proportion of clearance tasks assigned to mine detection dog teams to over 50% of the total.

For NQR detectors in particular, vehicle mounted operations may be more cost-effective. The large electric power demand (typically 30 kW) needed for NQR might be easier to supply on a vehicle platform, though the vehicle speed might be rather slow. The vehicle might be able clear vegetation and provide an NQR ground scan at the same time.

It must be noted that equipment purchases for demining programs have often not been made on the basis of conventional project financing calculations. Much of the equipment is donated to demining programs, rather than being purchased from their operating budgets. Purchasing criteria have been strongly influenced by non-technical factors. In Afghanistan, for instance, there might be resistance to spending a significant proportion of the demining program budget on equipment to increase the productivity of individual deminers. This is because a large number of Afghans depend on income from family members involved in the UN sponsored demining program. A single deminer may distribute his pay to 20 or more dependents, and the money is a significant stimulus to the Afghan economy and the Pakistan economy in the border areas where Afghan refugess live. Any further strain on the economies of these areas could cause further political instability.

However, recent changes to UN Standards for Mine Action (UN 2001) reflect a move towards greater emphasis on overall cost-effectiveness, taking all socio-economic factors into account. Improved modelling of socio-economic parameters (WB 2001) is permitting more quantitative decision making, even in countries like Afghanistan. This allows the additional economic benefits of faster demining to be taken into account (i.e. reduced refugee housing costs, earlier resumption of agriculture production etc.). The calculations in this paper cannot take the socio-economic benefits of faster clearance rates into account, but these benefits are likely to increase the affordable cost of faster detection technology.

Acknowledgements

The major part of the funding for this research came from the Night Vision and Electronic Sensors Directorate, US Army Fort Belvoir and private donations. The author acknowledges the contributions of many people to this work, particularly staff of the mine action centre for Afghanistan. Thanks are also due to his for support and many people in the demining community for helpful suggestions and comments, and the staff of the Hameed and Ali Research Centre for carrying out many field experiments to refine techniques described in this paper.

Reference

Trevelyan, J. P. (1998) Modelling Minefield Clearance Performance. Technical Report, Department of Mechanical and Materials Engineering, The University of Western Australia, Nedlands 6907.

UN 2001. Proposed Mine Action Standards. UN Mine Action Service, available from http://www.mineactionstandards.org/, hosted by James Madison University Humanitarian Demining Information Center.

WB 2001, Socioeconomic Impact of Mine Action in Afghanistan, Draft Report compiled by World Bank, available from UN Mine Action Centre for Afghanistan, Islamabad.

Technology and the Landmine Problem: Practical Aspects of Mine Clearance Operations

James Trevelyan
Department of Mechanical and Materials Engineering
The University of Western Australia
E-mail: James.Trevelyan@uwa.edu.au

When the enormous humanitarian consequences of widespread antipersonnel landmine contamination became apparent in the mid-1990s researchers in many countries tried to apply advanced technologies to make it easier to detect buried landmines. After seven years of intensive research and development efforts there are almost no practically useful results. One of the reasons is that most researchers appear to have only in a very superficial understanding of the practical aspects of mine clearance problems. It is tempting to imagine that minefields resemble a slightly rougher version of the neatly clipped lawns surrounding typical research laboratories whether they be in California, Bavaria, Tuscany, St Petersburg or Sydney. In this paper I will attempt to illustrate some of the many practical situations that pose technology problems very different to those that most researchers have imagined.

This paper is a written version of a slide presentation given at the Advanced Research Workshop. Each of the slides is accompanied by a written explanation and at the end of the paper there is a conclusion which draws each of the separate threads of this presentation together.

Slide 1+2

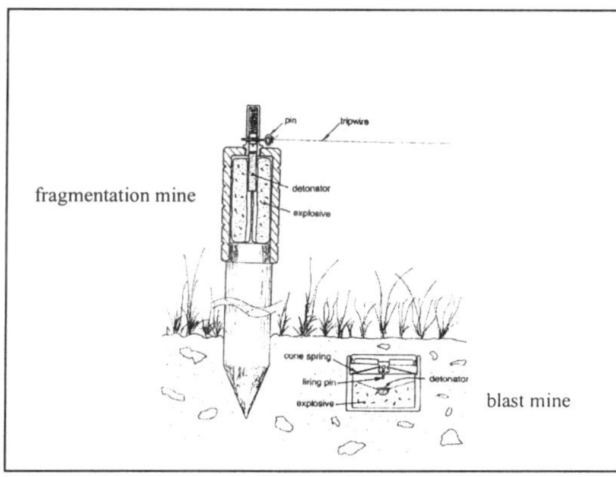

While most research has focused on the problem of the buried antipersonnel blast mine it is important to remember that several different kinds of mines are typically found during clearance operations. This slide shows two common kinds: on the right there is a simplified representation of a common type of blast mine buried just below the surface. On the left is a stake-mounted fragmentation mine: in essence it is a hand grenade mounted on a stake. Defensive minefields often employ both kinds together. The fragmentation mines provide the major defensive element. However, because they are relatively easy to see above the ground and the trip wires are often obvious, antipersonnel blast mines are laid around the fragmentation mines to deter any attempt to remove them.

Mine clearance operations may not start until years, even decades after the minefield was laid. The stakes will have fallen over and may have rotted, leaving the fragmentation mines half buried lying on their sides. The trip wires will either be lying on the ground half buried in rotting vegetation, or will have been lifted into the air as scrub and bushes grow in what was once a cleared area of land. If this happens the wires will run through the branches of the scrub and may pull the pins from the fragmentation mines as the branches sway in the wind.

Often, the fragmentation mines will not explode when this happens because of advanced corrosion in the fuse mechanism. However, the slightest mechanical disturbance can then allow the fuse to operate posing significant hazards for deminers.

Another kind of mine not illustrated here is the bounding fragmentation mine. This is equivalent in its effect to the stake mine shown in the picture but is buried under the ground. When its trip wire is pulled the

H. Schubert and A. Kuznetsov (eds.), Detection of Explosives and Landmines, 165–184.

mine propels itself a short distance into the air before exploding at waist height. Of course, fragmentation mines are usually easy to detect with metal detectors as they contain large amounts of metal.

Slide 3

Here we see several different kinds of blast mine. The PMN 1 (top left) is one of the most common blast mines in the world. It is extremely reliable: surveys in Cambodia have revealed over 90% in fully operational condition after 20 years. This kind of mine is extremely sensitive, particularly at the edges, and deminers tend to have more accidents with this kind of mine than others. It contains plenty of metal for modern detectors to find it at depths of up to 30 cm. The largest piece of metal is the stainless steel retaining ring that holds the rubber cap in place over the top of the mine. However, this ring is sometimes missing from the mines, reducing the metal detector signature substantially. The PMN 2 (top right) is another common antipersonnel blast mine with a much more complex design. Surveys in Cambodia have shown a much lower level of reliability. Further, because only the black rubber part is pressure sensitive, the rigid sections of the top casing provide more support for the soil over the top of the mine. Perhaps for this reason this mine seems to cause relatively few civilian casualties when the ground is dry and hard. The type 72 blast mine (lower right) is smaller and much more difficult to detect because there are only small metal components in it. This is the classic "plastic mine" at which researchers have aimed most of their efforts on detection technology.

Anti vehicle mines (lower left, photo: UN Mine Action Centre Sarajevo) are also a major issue in mine clearance operations. With few exceptions these mines have only small metal components and are difficult to detect with metal detectors. Since they have large amounts of explosive (7 - 10 kg) they are possibly easier to detect using advanced technologies such as nuclear methods and ground penetrating radar.

Slide 4

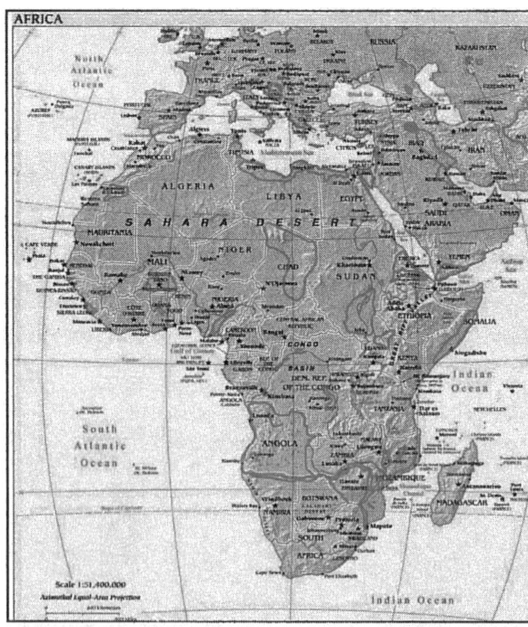

It is instructive to review the geographical spread of the landmine contamination problem. This slide shows the areas where mine contamination is a problem in Africa, the Middle East and parts of Europe. One-third of all the countries in the world are affected to some extent. Africa is perhaps the continent with the worst contamination problems.

At the same time, it is easy to overestimate the impact and extent of the mine contamination problem. Even in countries like Angola and Afghanistan, with some of the worst mine contamination in the world, people go about their daily lives living in and around areas of significant mine contamination. Often it is problems like the lack of good fresh water supplies, disease, few medical clinics and inadequate schools that are uppermost in the minds of people living with landmine problems. The mine problem has to be seen in perspective as only one of the issues affecting populations. For this reason, most countries affected by landmines will continue to rely on humanitarian aid from advanced countries to deal with the mine problem. They have scant resources of their own and they have to use those resources to alleviate their most urgent problems.

Slide 5

(Photo: MCPA)
Many people have suggested that large rollers could be pulled across a minefield, detonating all the mines making it safe for people to use. This picture, from Afghanistan, is a useful reminder of just how impractical ideas like this really are. The photograph was taken early in the mine clearance effort: Afghanistan deminers now have better protective equipment and tools. The main point about this photograph is the placement of the mine. In a recent conflicts involving militant factions rather than organized Armed Forces, all materials including mines were in short supply. Many mines were laid carefully to deny places where fighters could conceal themselves. Hence the placement of this mine beside the rocks where a sniper might be tempted to conceal himself. Mines were also laid in irrigation canals, alongside walls of ruined houses and many other places that offered cover and protection. All of these places pose major difficulties for any mechanical mine clearance device.

Slide 6

(Photo: ATC)
Here is another photograph from Afghanistan. We can see the entrance to a bunker used during the conflict. It is likely that the last fighters who abandoned this bunker would have left antipersonnel mines inside to deny other factions the use of the bunker. Mines would also have been laid in the trench system outside the bunker. Military fortifications like this are likely sites where mines are almost certain to have been laid. The effect of weather and erosion has been to bury the mines under layers of mud and rubble.

Slide 7

(Photo: ATC)
Here we see an anti-tank mine that has been exposed by excavating an area of ground where the mine was suspected to have been laid. This mine was probably located by mine detection dogs. The dogs do not indicate the exact location of the mine. Instead an area of approximately 4 to 6 sq m around the point indicated by the dog needs to be checked by manual deminers. Anti-

tank mines like this one contain almost no metal and are extremely difficult to find using metal detectors, especially in areas where there are many small metal fragments in the ground. In these cases the only method for finding the mines is slow and patient excavation as shown here.

In soft ground, thin probes could have been used to find a buried object like this. The hard ground and the stones in Afghanistan make this method impractical. Also, research has shown that probes are unlikely to be effective in finding mines deeper than 10 cm.

This kind of problem (anti tank mines) could be a useful application for ground penetrating radar: the targets are relatively large. It is important to realise that there are many distinct technical problems in demining. This is just one of many common problems faced on a routine basis by deminers every day.

Slide 8

(Photo: MCPA)
Not all mines are laid in the ground. The next photograph shows a fragmentation mine set up as a booby trap, hidden in the shadow of a niche, in a residential area of Afghanistan. Mines and hand grenades are often set up in this way and deminers need to be aware of all the tricks used by soldiers.

Slide 9

The main objective of mine clearance operations is to define areas of land that are free from mine contamination. Often, clearance priorities are dictated by the need for land. Any area that can be declared free of contamination without having to undertake complete manual clearance is obviously attractive.

An important aspect is the simple task of marking areas of ground. In this slide we see a Cambodian deminer marking the safe area in which he is squatting and preparing to mark a clearance lane. He has laid two wooden poles on the ground: the area between the poles will be checked with a metal detector. Marking the ground like this is an essential task that takes considerable time. In some countries where the ground is hard and stones are available, white and red painted stones are used instead of stakes and string.

Slide 10

Another task that has to precede any attempt to detect mines is clearance of vegetation. Metal detectors need to be close to the ground to work effectively. Therefore all vegetation must be removed. Here we see a deminer "dismantling" bushes. He needs to work with great care because there may be tripwires in the branches of the bushes. If he fails to notice a tripwire he could detonate a fragmentation mine nearby. The vegetation we see here has grown since the mines were laid: this represents about 15 years of growth from what was an open agricultural area before the conflict began. Tripwires originally near the ground can be lifted by vegetation as it grows.

Slide 11

Here we see a Schiebel AN 19 metal detector being tested. The deminers prepare a small area of ground and use a standard test piece to check the metal detector. In this area there are problems with soil mineralization: the detector confuses the soil with metal. This test is to check that the detector can be used.

The deficiencies of metal detectors have been well documented. Minefields typically contain large quantities of metal fragments, rubbish, discarded tin cans, cigarette packets and other domestic goods, bullets, and shell cases. All these are false alarms but each metal target has to be checked because it could be a mine. Deminers often find between five hundred and five thousand metal fragments for every mine they find. This means that demining is a hazardous and tedious occupation. Any technology improvement that can reduce the number of false alarms could be very useful.

Slide 12

Here we see a deminer using a thin probe to check the location marked by a deminer with a metal detector. The marked location is shown with a small white cone. The use of thin probes like this is falling out of favour because it is difficult to check to a depth of 20 cm -- a frequently used standard for manual mine clearance. Instead of using a probe, deminers are required to excavate the area around the marked location, searching to at least 20 cm below ground level, and removing any pieces of metal found in the process.

Once the piece of metal causing this "indication" has been found, the remaining area between the two wooden poles will be checked. When it has been completely checked, the two poles will be moved forward and the markings on each side of the deminer will also be moved. Then the process is repeated for the next small area of ground.

You may have noticed the paradox in this photograph: the deminer is working in a ploughed rice padi. In this part of Cambodia occupied by IDP's (internally displaced persons) people are desperate to grow their own food and occupy land despite the obvious risk of mines. Land is so valuable that they will take great risks to occupy and farm it in the hope that they can claim it eventually as their own. In these areas the original records of land tenure were lost and many of the families that used to live here disappeared during the war. Even though there are mines, land like this represents the chance for people to build their own livelihood and future security.

Slide 13

This photograph shows a soft drink store alongside the main road between Battambang (the second largest city of Cambodia) and Bangkok in Thailand. In the foreground there is a minefield sign. The soft drink store lies in a minefield and the worn track from the road shows that many people have stopped to buy soft drinks here. The houses also lie in the minefield and deminers were already at work removing mines from the rear of these houses. Interestingly, there are very few civilian casualties in this area during the dry season. The mines are almost all PMN 2's. The ground is hard baked clay reinforced with vigorous vegetation roots. During the dry season the ground seems to be safe to walk on. Heavy animals like bullocks are used to pull rice ploughs and mine incidents are rare. However, the local people know that as soon as the rain falls the ground will soften and the risk of mine accidents greatly increases. This is reflected in casualty rates that increase during the wet months.

Slide 14

(Photo: A. Smith)
Dogs have been used for mine detection in Afghanistan since 1989. They were in use in Europe by the end of the Second World War and have been used extensively by military forces since then. However, it was not until 1995 that mine detection dogs became an effective part of the Afghanistan mine clearance program. There were many difficult issues not the least of which were the cultural issues associated with the use of domesticated dogs in a Muslim country.

The dogs have proved to be extremely cost-effective in the areas where they can be used. Typically they achieve clearance at approximately one-quarter the cost of manual demining (approx US$0.15 - 0.22 per square metre, manual demining costs between $0.60 and $0.80 per square metre).

The major problem with mine detection dogs is that we do not understand how they find mines nor the limits on their effectiveness. Until 1998 the dogs demonstrated an exceptional record of reliability: once land had been checked by two dogs it was considered to be 100% safe. By the end of 1999 several accidents had been recorded in areas cleared by mine detection dogs. Similar concerns had also emerged in Bosnia and other countries. After a series of meetings involving leading dog experts and practitioners it has become clear that there is no clear scientific understanding to back up the use of mine detection dogs. The international community has set provisional standards for the use of dogs and has also initiated a major research effort to understand their limitations.

Slide 15

(Photo: D. Edwards)
Armoured mechanical excavators like the one shown in this photograph have proved to be extremely useful and versatile machines in Afghanistan mine clearance operations. In these residential areas, mines lie buried in the ruins of houses along with unexploded shells and rockets. As successive bombardments and erosion due to rain increased the depth of rubble, more mines were laid on top because fighting in these areas lasted for many years. This is extremely difficult work for deminers and the machines can save a great deal of time. Machine operation is observed by a second deminer protected in a bomb suit: he communicates with the driver by radio.

Because there are no written records of property ownership all the walls must be retained intact to avoid further social conflict. This means that mine clearance is extremely tedious and difficult. The machines can reach some of the areas that need to be cleared but most of this kind of residential area has to be cleared by hand. This means excavating in the rubble to a depth of up to 2 m.

The excavator is protected with 25 mm thick mild steel plates. The windows consist of 25 mm thick polycarbonate with a 3 mm thick polycarbonate outer skin. Most mine explosions caused by the excavator result in only minor scratching of the polycarbonate surface. The outer skin can be readily replaced at less cost than replacing the whole window.

Slide 16

Fighting in the former Yugoslavia between 1991 and 1995 resulted in an enormous area of mostly low-level mine and unexploded ordnance contamination. This map shows the approximate extent of this contamination. It is believed that between 50% and 70% of mined areas are known. This estimate is derived from accident statistics: slightly more than half of the recorded civilian accidents occur in areas known to contain mine contamination. Most of the contamination is concentrated along the confrontation lines that divided military factions during the fighting: river banks, buildings, abandoned industrial sites and forests. The total area of contamination is thought to be approximately 15,000 sq km of which only a tiny fraction has been cleared so far.

Slide 17

For most of the minefields being cleared in the Balkans the main issue is vegetation. Apart from the coastal regions demining is restricted to the warmer months when vegetation grows at an amazing rate. The mine clearance flail has turned out to be a very useful machine, not for clearing mines, but for clearing vegetation which can contain trip wires. Armoured flails resist damage due to trip wire activated mines and the occasional blast mine that might also be activated. Anti vehicle mines still pose a significant hazard: in areas where they occur special procedures need to be adopted. Armoured flails can withstand limited numbers of anti vehicle mine explosions provided that the mines are exploded by the flail machinery and not the wheels or tracks of the vehicle.

The key to successful use of flail machines has been the carefully planned combination of machines, mine detection dog units, and manual deminers. In this way the unit cost of demining has been reduced by two to three times. The use of commercial contractors in competition with one another has also helped, though the imposition and monitoring of strict quality control standards has been an essential part as well.

Slide 18

This is a photograph of a recreation park that has just being cleared using this method. Notice how high the vegetation has grown: this is approximately two weeks growth. The clearance cost was approximately $5/sq. m. Before clearance most of the vegetation was approximately 20 m high, consisting of thin scrubby trees that had grown up around the original ornamental trees.

Slide 19

The flail machine used for this job was the Belarty flail from Slovakia, shown in this slide which I took in a maintenance workshop nearby where it was being stripped down for a regular overhaul of the hydraulics system. The flail head, on the left, is mounted on a large extending boom carried by a turret on a Russian T55 tank chassis. This is a big machine: with a total weight of over 60 tonnes transport needs to be considered carefully. However, in Croatia the road system has been built for loads like this and there are several well-equipped maintenance facilities throughout the country. The small photograph on the right shows a single chain and weight from the flail unit. This particular machine is rated to withstand 40 anti vehicle mine blasts before a major overhaul is required. The extending boom arrangement of this machine provides much more flexibility than a conventional vehicle mounted flail: it can deal with steep embankments and uneven terrain. It is particularly suitable for the recreation park shown in the previous photograph. The local authorities had specified that there had to be no damage to the ornamental trees that were still growing in the park.

Slide 20

(Photo: Croatian Mine Action Centre)
Another machine devised for local demining requirements is this remotely operated mini flail known as the MV3. This machine is designed and built in Croatia. It has proved to be particularly useful for clearing around houses and gardens. These areas are important for resettling refugees and internally displaced people. Once again it is used for clearing vegetation: it is not intended to detonate mines. However, it has to withstand damage due to mine blasts: particularly bounding fragmentation mines.

Slide 21

The slide shows a recently cleared area around a culvert (a small bridge for a drainage channel): I am standing on the road just above the culvert itself. The drainage channels leading to the culvert were extensively mined because this was a major confrontation area between the two sides fighting in the area. Only the immediate area around the culvert has been cleared. The thick vegetation and scrub contains tripwire activated bounding fragmentation mines. It is expensive and slow going to demine this by hand and the ground is too uneven to attempt this using mechanical clearance. Only the area immediately around the culvert has been cleared to enable essential maintenance to be done on the culvert itself.

Slide 22

Although it does not look like it, this is a photograph of a house in a village, taken from the road outside. The house is completely obscured by high vegetation that has grown in the 8 years since fighting started in this area. Hidden in this vegetation there are steel fence posts and rusty domestic appliances that were taken from the house when it was looted. The appliances were too heavy to take away at the time so they were trashed to make sure they were of no value to anybody else. Impact marks of mortar shells on the roadway showed that this was an area where intense fighting occurred so mine clearance is needed before the houses can be repaired.

Slide 23

A ruined church in the same village being inspected to assess repair costs. A mortar impact can be clearly seen just in front of the two men. Limited clearance was carried out to enable the electricity supply to be restored. Following this, local people started to clear the area around the church themselves. They knew that at least six anti vehicle mines had been laid around the church but did not know where. Mine clearance operations often commence in areas that have already been cleared partly by local people: the local people give up when they think that they cannot handle the mines themselves. It is not uncommon to find significant amounts of unexploded ordnance collected by villagers and gardeners and stored in neat piles near their houses at great risk to themselves.

Slide 24

Here is a typical house in a typical Croatian or Bosnian village. How did it become like this? Usually an armed gang would appear in the village without warning, driving through and firing bursts of automatic weapon fire at the houses to wake up the inhabitants. All the people would be given 15 minutes to leave their houses taking whatever posessions they could carry and drive away with. Then, all the houses were systematically looted. All removable items of value were dragged out and placed in heaps outside to be collected later. Then, the houses were either burned or blown up. Sometimes mines were laid as well, but this did not usually occur unless extended fighting occurred in the area between military factions. From time to time, derelict houses would be looted again usually for firewood from windows, doors and flooring. More recently, abandoned houses have been looted once again for pipes and plumbing fittings even when they have been installed in cement. This is done by destitute people who are struggling to repair their own houses. Trees have grown up inside the house weakening the foundations.

Slide 25

Before minefields created

8 years after minefields created

These two pictures show the transformation of agricultural landscape caused by landmines. The first photograph shows the original condition of the landscape just north of the border between Croatia and Republika Serbska (the predominantly Serb part of Bosnia-Herzegovina). The second photograph was taken just south of the border where there had been extensive minefields for 8 years. This area was heavily fortified because there were no natural barriers preventing a military advance by Croatian forces.

Trees and vegetation, particularly thorn and prickle bushes have grown in the mined areas. The original drainage system in this area has deteriorated with the lack of maintenance so the ground is soft and boggy, and often flooded after periods of heavy rain. **It is extremely difficult terrain for demining operations yet, if** restored, could be valuable and highly productive agricultural land. Paradoxically, even if the land were cleared and restored to agricultural production, there would be no market for the produce given the current economic situation in Eastern Europe.

Slide 26

These two photographs were taken in 1997 near a factory complex near Karlovac, not far from Zagreb. The residential complex was trashed and looted before it became part of the confrontation lines between Croatian and Serb forces. Mines were laid in the piles of trash outside the buildings and the area has since become overgrown with vegetation.

Slide 27

(Photo: D. Busuladzic)
Here is a similar scene, this time in Sarajevo. Here we can see a deminer at work. The rate of progress is extremely slow. Each piece of trash has to be treated as if it were booby trapped. A hook is attached carefully then the trash is pulled from a safe distance. Once again, vegetation complicates the job. Armoured excavators could make this job much easier but they cannot be used in this location because the area was also laid with anti vehicle mines.

Slide 28

(Photo: D. Busuladzic)
A similar nearby scene, again in Sarajevo.

Slide 29

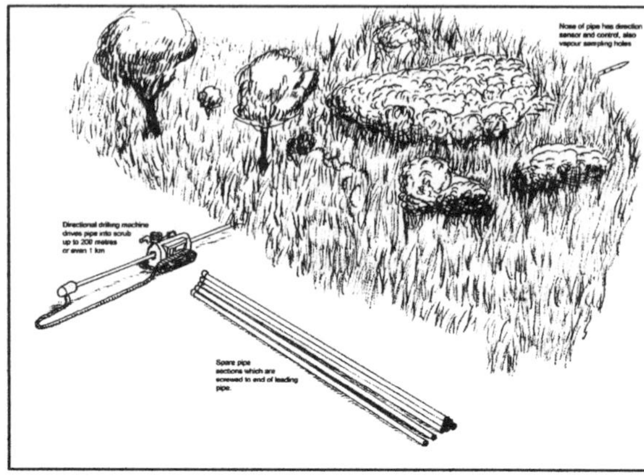

Here is one untested concept that could provide a better way to deal with many of these situations shown in the preceding photographs.

Mine and unexploded ordnance contamination is quite thin in many areas: it is not uncommon to find no mines in a typical demining task. The reason is that all the known minefields were cleared by the factions that laid them after the signing of the Dayton peace accords in 1995. However, the clearance effort was incomplete. Many of the minefields were too dangerous to clear without proper equipment and techniques. Even in those where clearance was accomplished it was seldom 100% effective. Therefore, all these areas need to be cleared properly but the residual contamination level is sometimes very low. Therefore, a lot of time and money could be saved by devising methods to confirm if there is any contamination before mine clearance starts. It is highly desirable that this be done without first having to clear the vegetation.

This diagram shows one possible way to achieve that. A directional drilling machine, in common use for installing underground power cables, is being used here to push a sampling tube through scrub and rubble to sample the air and dust near the ground surface. The air and dust is drawn through a filter pack at the rear of the machine. The filters are changed regularly. The filters are then taken to be examined by explosive detection dogs: a technique in regular use in South Africa. Sampling tubes would need to be inserted every 10 m or so. The aim of doing this would be to detect any trace of explosive contamination: if none were discovered the area could be classified as uncontaminated, or at least treated as a low-risk area.

Slide 30

Hundreds of kilometres of river banks and the mud banks in the rivers are also potentially contaminated with mines. Rivers were often dividing lines between military factions and the banks were heavily fortified. In this photograph we are standing in Croatia looking across at Republika Serbska in Bosnia. Somewhere near here several traps were created in the river bed when the water was at a low level. Each trap consisted of approximately 500 kilograms of dynamite with an arrangement to detonate it remotely. These traps were designed to deter attempts to cross the river with amphibious armoured vehicles. Now their location is unknown. No one knows whether the traps are still in place or whether the dynamite has been washed downstream.

The river often floods well above the level of the man shown standing in the picture. When this happens large amounts of floating wreckage and vegetation become trapped along the river banks. Many antipersonnel mines have been discovered in this waterborne material. This means that special precautions have to be taken to prevent this material from re contaminating cleared areas along the river banks.

At the time of writing there is a joint project funded by the European Commission to develop an underwater mine detection robot to help alleviate the problem of mines been washed down rivers. (BULLRUSH project)

Slide 31

(Photo: Colin King)
Cluster bombs posed a major difficulty for deminers working in Kosovo. They are extremely unstable if they fail to explode on impact. Some, such as the BLU 97 shown here, can explode as a result of the electromagnetic field from a metal detector.

Slide 31

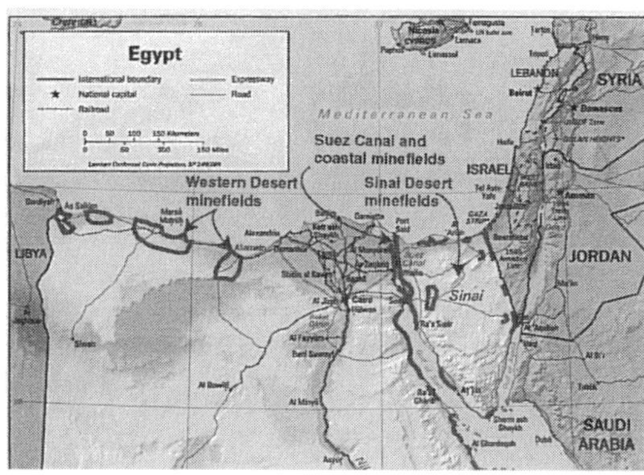

(Map: CIA web site, additional marking by author)
Egypt presents very different mine clearance problems. There are two principal areas of mine contamination. The area west of Cairo resulted from the conflict between 1939 and 1943 involving German, Italian, British and Commonwealth forces, the latter being assisted by Egyptian forces. The area of mine contamination extends well into Libya as well. East of Cairo the contamination resulted from conflicts with Israel in 1948, 1956, 1967, and 1971-1975.

Egyptian authorities are very cautious about releasing information on mine contamination. They are acutely sensitive to the impact this might have on their tourism industry that is the major source of foreign exchange income for the country. At the same time there have been misunderstandings between Egyptian authorities and mainly western countries that have been asked to assist with mine clearance operations. Western authorities accuse the Egyptians of inflating their statistics on mine contamination by including unexploded ordnance. However, this may result from the common use of the same Arabic word (Al Gham) to describe both items. In most countries unexploded ordnance is relatively safe to handle: the chance of accidental detonation is low. This means that mine and unexploded ordnance problems are often treated separately. In Egypt, particularly in the Western Desert, this is an artificial distinction because much of the unexploded ordnance is now in a very unstable condition because of its age. From a practical point of view both mines and unexploded ordnance need to be treated as equally dangerous.

Slide 32

Here we see a recently cleared area in the Western Desert near Al Alamein. Each red flag indicates a mine location. The mines are metal cased British anti-tank mines.

Drifting sand has covered them to a depth of up to 30 cm. Each mine has been excavated and lifted to the surface. They will be carefully transported to a disposal pit nearby.

The mines were located using metal detectors. The amount of metal contamination in this area is quite low as it was far behind the areas where fighting took place.

Slide 33

This shows a close-up view of one of the mines. The corrosion is so advanced that we can see the explosive inside the casing. This part of Egypt experiences prevailing winds that blow off the Mediterranean Sea carrying significant quantities of salt. Even though there is not much rainfall most metal items have advanced corrosion after 50 years of exposure.

Slide 34

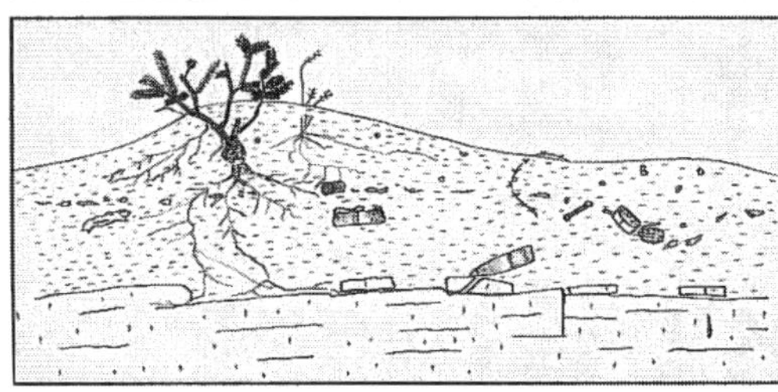

The drawing shows what I have been able to deduce from my conversations with Egyptian authorities and my visits to the area. The ground level at the time of fighting in the 1940's is littered with metal fragments: rubbish, spent ammunition, fragments of shells and vehicles hit by shells and barbed wire. Unexploded shells lie in the sand beneath this layer, possibly adjacent to the limestone bedrock between 30 cm and 2 metres below the surface. Small plants have grown, attracting drifting sand and raising the ground level around them. In some areas the wind has exposed formerly buried items. If it were not for the presence of large unexploded ordnance (up to 500 kg) it might be possible to use specially adapted vehicles to carry out mine clearance operations by separating the sand and collecting all the metal material.

Slide 35

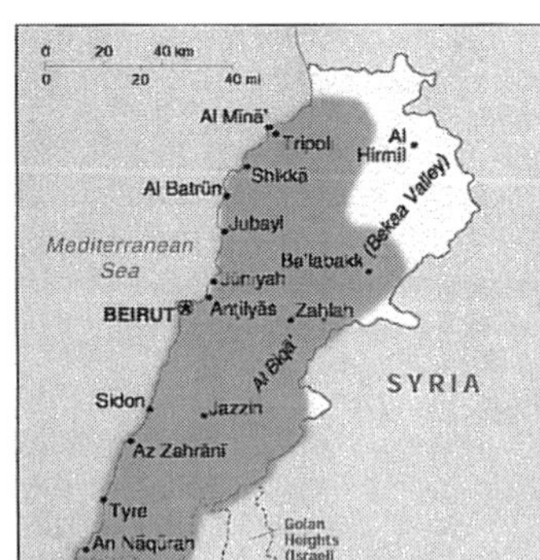

Until recently only the southern area of Lebanon was considered to have a mine contamination problem where Israeli forces constructed extensive minefields around their defensive positions. Isolated problems were evident in the rest of the country, particularly around the capital Beirut. The reason why the contamination in the rest of the country lay hidden for some time can only be understood in the context of Lebanese politics. The country is ruled by a Christian minority government. Most of the mine problems affect marginalised Muslim populations with scant health-care resources. The extent of mine contamination only became apparent when government health authorities investigated higher than average consumption of artificial legs and arms.

In most countries mine contamination is often discovered by agencies like the Red Cross when they have to provide support for mine victims. However, with the extensive civil war in Lebanon and frequent kidnappings, outside humanitarian agencies have been completely absent from many areas.

A further complication is that 60 percent of Lebanon is still occupied and tightly controlled by the Syrian army.

As in Croatia and Bosnia, most areas have light contamination but the problem is widespread, particularly as a result of the use of cluster bombs in the Bekaa Valley.

Slide 36

These photographs show aspects of a minefield overlooking Beirut. This was a key military position during the civil war. Now it is a keenly sought after area for prestige housing development: luxurious houses are under construction immediately adjacent to the minefield being cleared. The area was subjected to intense artillery bombardment including 16 inch shells from the US Navy. The resources of the small units of the Lebanese army specialising in ordnance disposal are thinly stretched. They spend much of their time on calls to construction sites where they are asked to deal with unexploded shells and mortar bombs. Financial considerations mean that construction sites receive higher priority than mine clearance tasks in most areas. It should be remembered that much of the ammunition used in low intensity conflicts has a relatively high failure rate. In Yugoslavia, for example, it is estimated that only one-third of the ammunition used in the conflict exploded on impact as intended. This means that large amounts of unexploded ordnance remain after conflict has finished.

The photographs show that the mine clearance task itself is not easy. The terrain is very rough limestone and metal detectors cannot be used because there is so much metal contamination from the artillery bombardments.

Slide 37

Here we move to Jordan. We are looking at a hole about 60 cm deep. Three American-made M14 antipersonnel mines have been removed from the bottom of the hole. It is not easy to understand how they came to be there.

This minefield is in the Jordan River valley. It is a high priority for clearance because the land is the most productive agricultural land in Jordan.

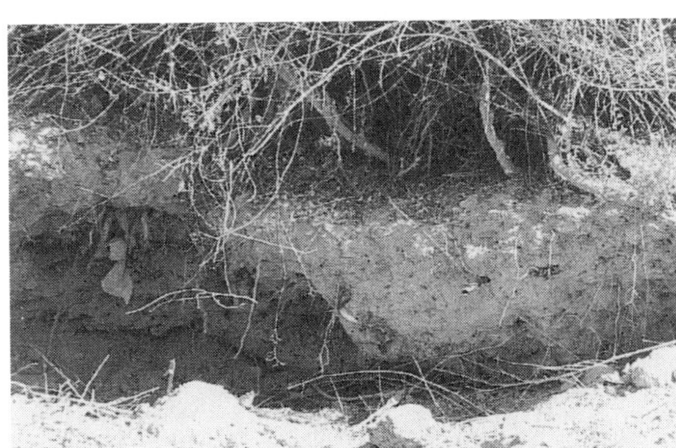

The photograph shows irrigation piping that was in place when the mines were laid after the 1967 war. Interestingly, the irrigation piping is about 50 cm above where the mines were found.

The mines were laid according to British patterns. They were laid with great accuracy in a pattern marked by steel picket posts: each anti-tank mine was surrounded with three antipersonnel mines. The location of the minefield is accurately known: the steel picket posts are still in place. Yet, this minefield has proved to be extremely expensive to clear. Further, there are many more minefields like this in the Jordan valley. Once the mines were in place vegetation grew uncontrolled in the minefield areas. On the occasions when storms cause flooding in the valley, the vegetation trapped mud and silt carried by the water, increasing the ground level. In this way, the mines were gradually buried to their present depth up to 1 m below ground level. Around 80% of the mines are almost exactly where they were laid. The remainder have moved.

The minefield is being cleared by a painstaking manual method. Deminers work along the line of the original minefield carefully probing the ground where the mines ought to have been laid. For much of the time they have to dismantle vegetation including medium sized trees, and carefully remove the root structures underground. They work deeper in layers until they finally find the anti-tank mines. Once the anti-tank mines have been found they can locate the three antipersonnel mines at known distances away from each anti tank mine. If these mines are not within 20 - 50 cm of their specified locations they presume them to have been washed away.

Given the number of men involved in clearing this minefield and the rate of progress, using local labour rates, I calculated the clearance cost at approximately $250 per sq m or about 200 times the normal clearance cost for manual demining! Given that this minefield was laid in a precise military fashion and its location is exactly known with exact numbers of mines laid, this instance serves as a useful reminder to rebut the common argument that "minefields laid by organised military forces can easily be cleared".

Conclusions

New Technologies?

The photographs and case studies in the first part of this paper represent just a sample of all the different problems faced by mine clearance organisations around the world. What can we learn from these case studies about the applicability of some of the new technologies that have been proposed?

Nearly all the research and development in the last seven years has been focused on the problem of detecting plastic mines with zero or minimum metal content. Yet the total proportion of the effort in mine clearance operations that could be saved by using a man portable "mine detector" (rather than a "metal detector") is not as large as one might expect. In manual clearance operations in some countries where there is little or no vegetation (such as Egypt or Afghanistan for example) the potential time saving is significant. Yet, even in these situations, the work involved in marking the ground and searching for targets is still significant. In countries like Cambodia or Croatia clearing vegetation to provide access for detectors takes as much as 75% of the total effort. In these situations it may be hard to justify expensive detection equipment.

By far the greatest common need expressed by demining organisations is a technology that will confirm the absence of mines or other explosive devices in areas that are suspected to be minefields, without having to clear vegetation, rubble, or in other ways having to work the ground in a potentially hazardous situation. Unfortunately very little work has been carried out with this objective in mind.

Nearly all the work associated with new technologies aims to replace current metal detector technology with some more advanced means of sensing. The real underlying need for this is to reduce the number of false alarms associated with using metal detectors.

Ground penetrating radar (GPR) in combination with a metal detector has been the favoured technology since 1995. GPR only works well in ground that is slightly damp. In wet ground too much radiation is absorbed. In dry ground there is little difference in the dielectric constant between the mine and the ground so the reflected energy is substantially reduced. Even in damp conditions, metal fragments, stones, disturbances in the ground caused by borrowing animals and insects, rubbish and large roots also provide false alarms for the radar. No one has yet been able to demonstrate any improvement in cost effectiveness through the use of radar.

Detection of explosive vapour has long been thought to be the ultimate method for detecting explosive contamination from any source. However, even the resources of the US Department of Defense have not been sufficient to resolve the technical problems (refer DARPA "Dog's Nose" project). It is possible to detect vapours from most mines containing TNT under most conditions but not 100% of the time and certainly not fast enough to match manual demining rates. Mines that contain explosives with very low vapour pressures (PETN, RDX, HMX) cannot be detected this way.

Nuclear quadrapole resonance is currently being researched intensively on behalf of the US Department of Defence by Quantum Magnetics Inc. and by the British Defence Experimental Research Agency (DERA). A man-portable prototype is expected by the US in 2006. While this technology works reasonably well for advanced explosive materials, TNT turns out to be particularly difficult because of the low intensity signals arising from it. Further, degraded TNT in old ammunition has even more complex low intensity signals making detection particularly difficult. Several other nuclear techniques have also been tested and research is continuing in Europe and USA.

Most of the new technologies are seen as "secondary" or "confirmation" sensors. This term means that they are intended as a means of discriminating a target located with a metal detector to decide whether it is likely to contain explosive or not. Much of this research has been conducted under the misapprehension that it takes a long time for deminers to decide whether a metal target is a mine or not (30 minutes or more). A further justification is that the process of excavating a metal target is intrinsically hazardous.

Recent analysis that we have completed in our own research shows that the time needed to discriminate a target by manual excavation is only about 90 seconds on average. This analysis is based on records maintained by the Afghanistan mine clearance organisations between 1995 and 2000 during which time about 2500 minefields were cleared. Further, while accident rates used to be unacceptably high, they have been reduced to levels comparable with driving motor vehicles.

Given the research results obtained so far, there is unlikely to be a significant technological development that would significantly improve cost effectiveness in the next five to ten years. However, our own recent research suggests a further reason why ultimate success is likely to be limited.

Slide 39

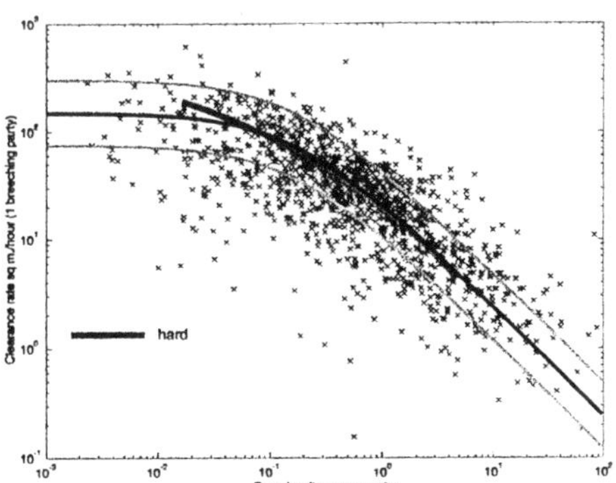

This graph shows a logarithmic plot of the minefield area clearance rate in sq m per hour (vertical axis) with the fragment density per sq m (horizontal axis) for about 700 minefields in Afghanistan with hard ground conditions, cleared between 1995 and 2000 using manual methods and a single demining team. The three curves represent a simplified model of clearance rate: a constant component of 150 sq m per hour with a fixed penalty of 2.5 min for every fragment found. (The two lighter lines represent constant clearance rates of 75 and 300 sq m per hour.) While there is considerable scatter, careful statistical analysis supports this conclusion with a high degree of reliability.

Slide 40

This photograph shows measurements been made on a fragment excavation in simulated demining tests in Islamabad, Pakistan. These tests were conducted to test the statistical conclusions from our study of Afghanistan mine clearance statistics. They showed that it takes around 30 to 60 seconds from the first moment a deminer hears a target indication from a metal detector to the start of the excavation operation. This includes the time required to mark the location on the ground. Subsequent excavation takes 90 seconds on average providing excellent support for the conclusions from statistical study.

Possible improvements may be limited

Let us assume, optimistically, that some new technology becomes available that can act as a confirmation sensor for metal detector. Let us assume that it takes 15 seconds for this device to decide whether a metal target contains explosive or not. We can use the results of our statistical analysis to predict the time saving that would result from using this detector for demining operations. With a fragment density of one per sq m which is average for Afghanistan, assuming that the false alarm rate is reduced by 99%, the improvement in the area clearance rate is only 70%. Given that the cost of employing one deminer for 12 months is around US$10,000 (higher rates apply in some countries: most of the cost is not the deminers salary as this figure includes all the activities and costs that support a deminer in the field), improving his productivity by 70% could save about US$7000 per year. In the country with significant vegetation problems the improvement in productivity is much less: approximately 30%. This means that one cannot justify investment in significant technology on economic grounds alone.

This simplified analysis does not take into account the improvement in economic return from land which could be cleared quicker than present rates: if the land is put into productive use sooner we can offset increased clearance costs by the increased return from using that land. However, the only land that has sufficiently high economic value for this factor to make a significant impact is in urban areas where the technical problems are far greater.

Another undesirable side effect of investing so much in advanced technologies is that we have not made desirable improvements in metal detection technology. This photograph shows the Australian Minelab F1A4 detector that we tested in Islamabad. It has been very successful in overcoming the problem of mineralized ground in Cambodia and several other countries. It was based on a design for a gold detector in Australia where the ground conditions are extremely difficult for metal detectors. This development is practically the only improvement in metal detection technology that has been made in the last ten years. Yet, recent tests conducted in several countries have revealed that there are significant performance problems with current generation metal detectors. Most of the companies producing metal detectors are small and do not have significant funding for research and development. Funding has only been provided for technologies other than metal detection. Therefore, it is likely that the present generation metal detectors could be significantly improved.

Mine Detection Dogs -- Performance Problems

Before concluding it is worth remembering that the most effective clearance method in use today involves mine detection dogs.

The University of Western Australia, along with several other organisations, was about to take part in a major scientific trial of mine detection dogs in Afghanistan: the trial was due to start on the 12th September. Given the present situation in Afghanistan the trial is now not expected to begin until at least 2003.

The problem here is that we do not understand how dogs manage to find mines, and therefore, we do not understand why the dogs occasionally miss them. In a few cases the dogs have missed an unacceptably large proportion of them. Large areas of land have been certified as safe with a method that we do not fully understand, so we cannot be completely confident of the results.

In the meantime, research continues on dog capabilities in the USA, Germany, Sweden and at the Geneva International Centre for Humanitarian Demining, where UWA is involved.

Risk Reduction or Risk Elimination?

We need to remember here that recent tests (ITC 2001) with metal detectors have underlined, once again, the technical weaknesses of current detection technology. Several mines were missed by the majority of detectors even though the mines were well within the specified clearance limits used in most mine clearance operations.

The essential aspect of mine clearance is that it is a risk-reduction process, not a risk elimination process. However, attempts to introduce a risk-reduction framework into mine clearance have not yet been successful: the industry does not want to face the difficult task of educating sponsors about the concept of risk reduction. The unfortunate consequence of this reluctance is that alternative approaches to the mine clearance problem cannot be appreciated, so many low cost approaches are currently being ignored. Two examples will be sufficient to illustrate this:

a) Use of Machinery without Manual Demining Follow-up. It is widely accepted that machines, by themselves, cannot provide mine clearance to the required standards (as currently framed). However, when the initial contamination level is low, mechanical "ground treatment" may be sufficient to reduce the risk to an acceptable level.

b) Bypassing Mine Clearance. If adequate protection can be provided to operators, land may be productively farmed using entirely mechanized approaches. Anti-personnel mines can damage farm machinery, but the damage is comparable with normal farm incidents (e.g. collisions with large rocks, tree stumps etc.). By bypassing mine clearance and farming the land in this way, the long-term return may be considerably greater.

Sources of Reference Material

Rather than give a complete list of references, I have provided suggestions for further reading and sources for reference material.

All of the material on minefield situations has been drawn from the UWA Demining Research Project web site: http://www.mech.uwa.edu.au/jpt/demining/. The site is fully indexed and contains a wealth of background material, with detailed reports on each country. UWA research on protective equipment, quality measurement methods and other issues is also fully reported with peer-reviewed technical reports and papers.

The EUDEM and DETEC projects coordinated by Dr. Claudio Bruschini provide excellent reference material on detection technologies under development in different countries, particularly in Europe. Links to these projects are provided from the "International Links" page of the UWA web site.

The ARIS web site, hosted by the Joint Research Centre of the European Commission at Ispra in Italy, provides another comprehensive resource for demining technologies researchers. In addition, it also provides the recent ITC Report on Metal Detector Trials completed in 2001.

The Journal of Mine Action available from the James Madison University web site provides magazine style articles on mine action (including mine clearance) and a smaller number of fully referenced research articles. The magazine can be ordered in hardcopy form: it is currently distributed free of charge to demining organizations.

The Canadian Centre for Mine Action Technologies (CCMAT) provides useful information on demining technologies. The UWA web site "International Links" provides a link to their web site.

The principal reference source for this paper has been a series of visits to mine-affected regions and demining operations. Visits like this are the ultimate reference source needed for demining research. No amount of reading can provide an equivalent understanding of the problems that have to be solved: it can only provide a better preparation for a field visit.

Acknowledgments

The major part of the funding for this research came from the Night Vision and Electronic Sensors Directorate, US Army Fort Belvoir and substantial private donations. The author acknowledges the contributions of many people to this work, particularly staff of the Mine Action Centre for Afghanistan, Cambodian Mine Action Centre, Bosnia Herzegovina Mine Action Centre, Geneva International Centre for Humanitarian Demining, Croation Mine Action Centre, Lebanese Army, Royal Jordanian Engineers, and the Egyptian Military Engineering Organization. Thanks are also due to many people in the demining community for helpful suggestions and comments, and the staff of the Hameed and Ali Research Centre for carrying out many field experiments to refine techniques described in this paper.

THE HIGH EFFICIENCY COMPLEX FOR DETECTION OF LANDMINES

A.I.KAREV, V.G.RAEVSKY
Lebedev Physical Institute, Russian Academy of Sciences
Leninsky prospect 53, 117924 Moscow, GSP-1, Russia
J.A.KONYAEV, A.S.RUMYANTSEV
JSC Central Design Bureau "Almaz"
Leningradsky prospect 80, 125178 Moscow, Russia
R.R.ILIUTSCHENKO
15th Central Scientific Research Institute,
Ministry of Defense of Russian Federation
143430 Nahabino, Krasnogorsk, Moscow region, Russia

1. Introduction

As a result of local wars and terrorist action about 100-120 millions of landmines have been laid on territories of 64 countries. Every week 600-800 people become victims of the mines, most of them being women and children. Besides, these mine-contaminated areas are excluded from the local economy. This is the reason why the Unite Nations have included humanitarian demining into the list of most important international problems. The UN has also issued Standards for humanitarian mine clearance; the main requirement is: demining 20 cm deep with 99.6% probability [UN (1996)].

Unfortunately, modern methods used for demining do not such efficiency. As a result, the main devices used for the humanitarian demining are a portable metal detector and a pioneer probe. This results in a high risk for personnel and a slow rate of demining: a man can clear only 20-50 m^2 per day. For example, to clear the territory of former Yugoslavia by current hand method 150-300 years are needed.

One of the possible ways to overcome this problem is to use robotic technology that is based on new highly effective physical methods for direct explosive detection. The design of the mobile complex satisfying these requirements is discussed bellow.

2. Gamma-activation method for mine detection

The gamma-activation method for explosive detection was first proposed by Nobel Prize winner L. Alvarez, and later then this method was verified in work [Trower (1993)]. The basis for the method is presence of high concentrations of nitrogen and/or carbon in most of the explosives. The method relies on detection of decay products of short-living isotopes boron-12 and nitrogen-12. These isotopes with half-life 20.2 and 11.0 ms are created as a result of the following photonuclear reactions:

H. Schubert and A. Kuznetsov (eds.), Detection of Explosives and Landmines, 185–193.

$$\gamma + {}^{14}N \rightarrow {}^{12}B + p + \qquad \text{(threshold } E_\gamma\text{=24 MeV)}$$

$$\quad \longrightarrow {}^{12}C + e^- + \nu_e \; (t_{1/2}\text{=20.2 ms})$$

$$\gamma + {}^{14}N \rightarrow {}^{12}N + n + n \qquad \text{(threshold } E_\gamma\text{=31 MeV)}$$

$$\quad \longrightarrow {}^{12}C + e^+ + \nu_e \; (t_{1/2}\text{=11.0 ms})$$

$$\gamma + {}^{13}C \rightarrow {}^{12}B + p \qquad \text{(threshold } E_\gamma\text{=17 MeV)}$$

$$\quad \longrightarrow {}^{12}C + e^- + \nu_e \; (t_{1/2}\text{=20.2 ms})$$

The isotopes ^{12}B and ^{12}N are β-active and emit electrons and positrons with energy up to 13 MeV and 17 MeV respectively. When these charged particles move into a matter, they produce gamma-quanta. These gamma-quanta, as well as electrons and positrons, may be detected by a particle detector to form a useful signal.

When a gamma ray beam with energy about 30-70 MeV irradiates the soil, a lot of non-stable nuclei are produced. In principle, these isotopes can create background signals. However, most of these isotopes have a very short half-life (less than 1 ms) and their false signals may be easy selected if the pulsed mode of the irradiation is used and a counting device is switched on a few ms after the end of the impulse. Other isotopes have a too long half-life (more then 100 ms) and so the flux of the false signals from the isotopes will have little intensity in the measurement time window. Besides, the decay products from these isotopes have low energy – less than 4 MeV – and so the background may be reduced by means of energy discrimination in the counting device.

There are about 150 unstable nuclei having half-life from 1 up to 20 ms but in our case these isotopes can only be born in the soil as a result of very exotic reactions with very small cross sections.

In reality, there are only two reactions on oxygen:

$$\gamma + {}_8O^{16} \rightarrow {}_8O^{13} + 3n$$

$$\quad \longrightarrow {}_7N^{13} + e^+ + \nu_e \; (t_{1/2} = 8.6 \text{ ms})$$

$$\gamma + {}_8O^{16} \rightarrow {}_8B^{13} + 3p$$

$$\quad \longrightarrow {}_7C^{13} + e^+ + \nu_e \; (t_{1/2} = 17.4 \text{ ms}),$$

which can increase the level of the background, because the energy of the decay positron and the electron are 17.8 MeV and 13.4 MeV respectively. But the energy thresholds of these reactions are 52 MeV and 41.7 MeV, and so if the energy of irradiating gamma-quanta is less than these values, there will be not "false" signals.

Summarizing the above, if we irradiate the soil with a pulsed beam of 32-42 MeV X-rays, and the particle detector is switched on after a few ms after the end of the pulse , then in 10-20 ms we can get a useful signal from a explosive with high contrast.

The selectivity of the explosive detector may be increased if measurement of the outgoing radiation from the soil is done twice with different values of maximum X-ray energy in the irradiating beam. In this case we can separate the carbon and nitrogen signals. In fact, if during the first irradiation the maximum energy of the gamma-quanta is 40 MeV, we can get the sum of signals from carbon and nitrogen. If during the second irradiation the maximum energy of X-rays is 18-23 MeV, we will get the carbon signal only from the reaction $^{13}C(\gamma,p)^{12}B$. By means of this simple procedure we are able to obtain "pure" nitrogen and carbon signals and thus we can in principle determine the N/C elemental ratio in the irradiated area.

3. Computer simulation of the mine detection

To obtain the main parameters of the mine detector complex and to find optimal characteristics of the detector units we did the computer simulation of the detection process. Using a Monte-Carlo code with built-in database of isotopes we observed the history of each act of explosive detection starting with the emission of a gamma-quantum from the target (see Figure 1) up to counting procedure in the particle detector. Interactions of the accelerated electrons and gamma-quanta with the target, air, soil and explosive material, as well as the interactions of the decay products with the matter were taken in account.

The model of a mine detector shown in Figure 1 had the following parameters. The electron beam was incident on the 1 mm thick tungsten plate, and the particle detector had dimensions of the sensitive area 100×100 cm^2. The distance between the detector and the soil was 150 cm.

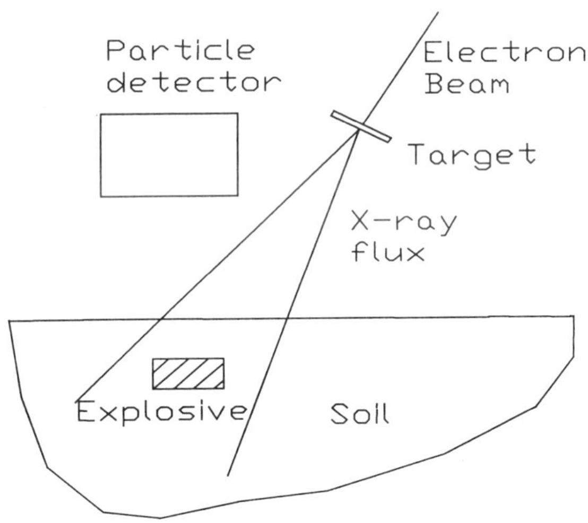

Figure 1. The general scheme of the model.

At first step of the simulation we found the optimal value of the electron beam energy. On one hand, the quantity of gamma-quanta, which produce useful signals, is increased when the electron energy is increased, and so the high energy of the electron beam is preferable to obtain the intensive useful signal. On the other hand, we have found that if X-ray energy is higher than 41.7 MeV, we can get false signals from oxygen. However the shape of the bremsstrahlung spectrum allows one to find the value of the electron beam energy when the level of the false signals is low. We calculated the rate of false signals from water and soil when these materials with high concentration of the oxygen are irradiated by an X-ray beam produced by the electrons with different energies. Some results of the simulation are shown in Table 1. The data were obtained for 50 mA electron beam current with 6 μsec pulse length, and X-ray tracks are followed up to 1 m deep. The counting process duration was 18 ms, and it was started 1 ms after the end of the irradiation pulse.

Table 1. The average level of oxygen false signals for electron beams with different energies.

Photo-nuclear reactions	Energy of the electron beam					
	WATER ($\rho = 1$ g/cm^3)			SOIL ($\rho = 1,6$ g/cm^3)		
	65 MeV	60 MeV	55 MeV	65 MeV	60 MeV	55 MeV
$O^{16}(\gamma,3n)O^{13}$ $O^{16}(\gamma,3p)B^{13}$	260	83	0,31	142	45	0,17

It can be seen that when the electron energy is 55 MeV the level of the false signals is negligibly small. So this value of energy of the electron beam was chosen for the explosive detector.

The results of the explosive detection simulation are shown in Figure 2. The geometry, the electron beam pulse, the target, the particle detector, and the counting process parameters of the model were given above. The families of curves presented in Figure 2 show the dependence of level of useful signals from TNT explosive with different weights on the depth for a clay soil and water.

The presented data indicate that the 40 g explosive can be detected in the soil at depths down to 30-35 cm and in water at depths down to 55-60 cm. A explosive ordnance with 125 g charge can be seen at depth 40 cm in the soil and at the depth 70 cm in the water.

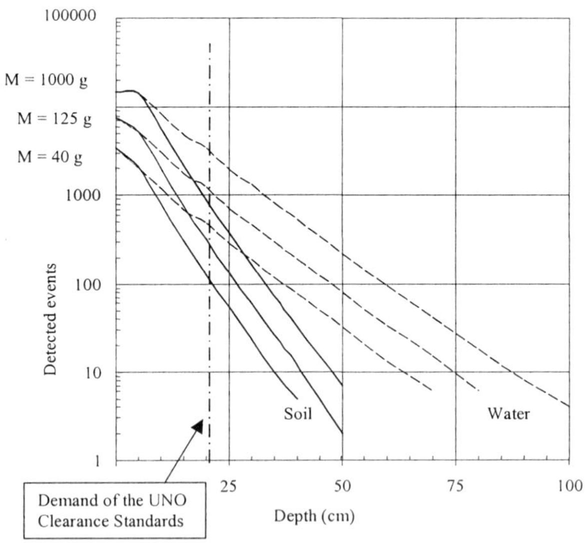

Figure 2. Dependence of the number of signals in the detector on the depth for TNT with different weight M in various absorbers – water and soil ($\rho = 1.6$ g/cm^3).

The probability of the explosive detection is determined by the level of background signals. From our estimates the level of the background from cosmic rays, air, and soil nitrogen as well as carbon is about 10 events per one measurement cycle. Comparing this value to the useful signal, one can see that the probability of detection of 40 g explosive is better than 99.6% down to 30 cm depth in the soil. In water the 40 g explosives may be detected with 99.6% probability down to 50 cm depth. Nitrogen and carbon from fertilized soil covered with grass and shrubbery increase the background and the depth of effective determination of 40 g explosive in the soil is decreased down to 25 cm. But even in this case the parameters of the mine detector satisfy the UN clearance standard.

Table 2. Number of useful signals from different kinds of explosives in soil.

Explosive	N	PETN	Hexogene	Octogene	Tetryl
M = 40g depth = 5 cm	2070	2258	4780	4937	3167
M = 125g depth = 20 cm	319	420	893	922	589

Note that the TNT is not the best explosive for detection as shown in Table 2. It means that the sensitivity and the detection reliability of the mine detector are better for other explosives and explosive blends (Composition B, Composition C-4, Semtex-H, etc).

4. Engineering aspect of the complex

The basis of the landmine detection complex is the accelerator-detector plant shown on Figure 2. The plant consists of the electron accelerator, the bremsstrahlung target, the bending and the scanning magnets, X-ray monitor and the particle detector with the counting device. We are planning to use the 55 MeV pulsed race-track microtron as an accelerator for this device. The Lebedev Physical Institute has practical experience in designing light, efficient, reliable and small machines [Belovintsev (1987), Belovintsev (1994)].

The main parameters of the 55 MeV race-trace microtron are listed in Table 3.

Table 3. The main parameters of the race-track microtron.

Output energy (MeV)	24-55
Output current (pulsed) (mA)	45-55
Pulse length (μsec)	6
Pulse repetition frequency (Hz)	50
Weight (kg)	2000
Dimensions (mm)	2000×600×500

The electron beam from the gun is injected into the accelerating structure and its energy is increased by a ΔE value. Then the accelerated electrons are directed back into the accelerating structure by means of two end magnets M1 and M2 to receive the energy gain ΔE again. So the beam energy at a given orbit is simply the number of the orbit multiplied by ΔE. To obtain the double measurement mode in the mine detector the beam may be extracted with different energies by ME1 or ME2 magnets. The source of RF power for the accelerating structure is a pulsed multi-beam klystron.

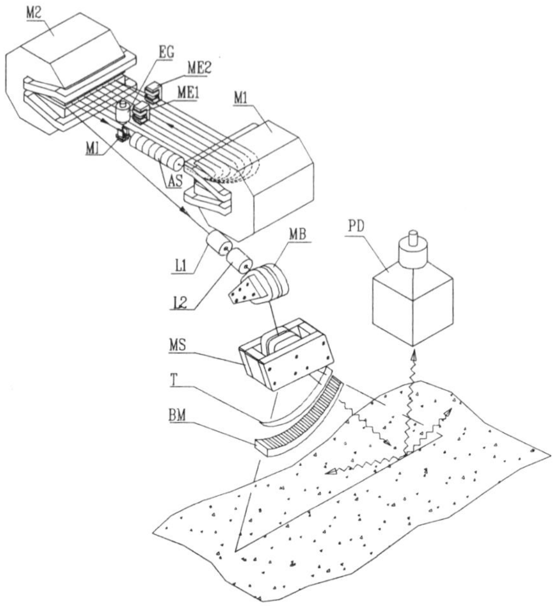

Figure 3. The general view of the accelerator-detector plant.

AS- accelerating structure, M1 and M2 - end magnets, EG - electron gun, MI - injecting magnet, ME1 and ME2 - extracting magnets, L1 and L2 - lens, MB and MS - bending and scanning magnets, T - target, BM - X-ray beam monitor, PD - particle detector.

The accelerated electron beam is focused by lenses L1 and L2 (see Figure 3), and magnet MB bends it in the direction of the soil. The scanning magnet MS with variable magnetic field turns the beam direction to the target. In the target the electron energy is converted to X-ray beam which preserves the direction of the electron beam. In this way, a line on the soil can be scanned. The position of the irradiated zone on the soil can be calculated from value of the current in the scanning magnet coil or can be obtained by means of the X-ray monitor, which is a multi-wire ionization chamber.

The landmine detection complex consists of the two units – remotely controlled moving explosive detector (MED) and a control cabin placed on a truck. The cabin is connected with MED by information channel and two-person team controls the MED operation. The main characteristics of the complex are listed in Table 4 and the general view of the MED mounted on a tank chassis is shown in Figure 4.

The electron accelerator, klystron, high voltage source and the analog and digital devices are included into the dust-proof container placed on the tank back. The bending and scanning magnets as well as the target, X-ray monitor and local radiation shielding are mounted into the radiating head. The particle detectors supported by mechanical arms may be set either under the soil (operating position) or on the tank back (run-in position). Besides there are the modules with AC source, water cooling system and the hydraulic crane.

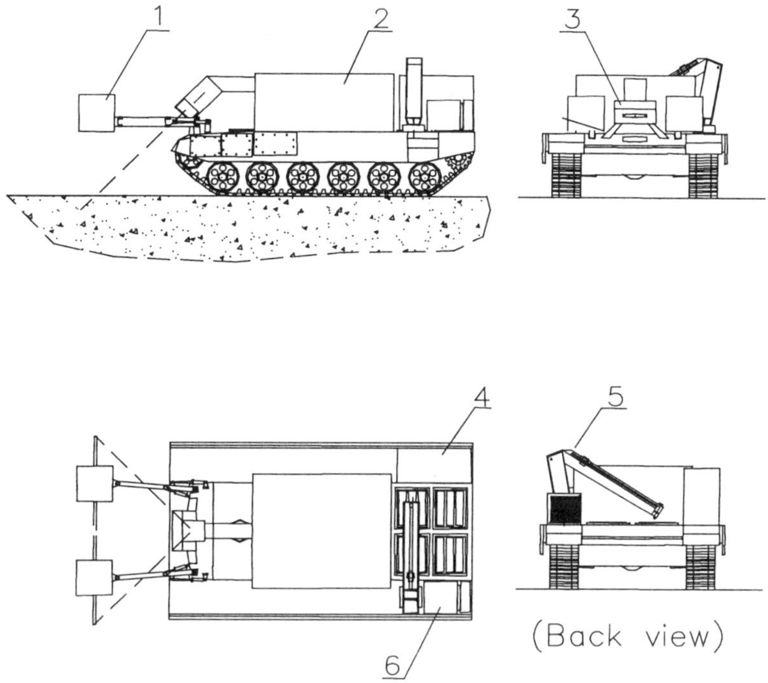

Figure 4. The MED on the tank chassis. 1-particle detector, 2- dust-proof container, 3 - radiating head, 4 - AC-source, 5 - hydraulic crane, 6 - water-cooling system.

Speed of the inspection	Up to 1500 m^2/h
Width of the inspected zone	4 m
Detected explosive mass (99.6% probability)	Depth:
In soil	
40 g	25 cm
125 g	30 cm
1000 g	35 cm
In water	
40 g	40 cm
125 g	50 cm
1000 g	55 cm

Table 4. The main parameters of the complex.

The operation of the complex will proceed in the following way. The MED controlled from the remote cabin moves with small velocity examining 4 m wide band of soil surface. When an explosive is found the MED is stopped, the location of the explosive is marked with paint and all information about the object including its coordinates is recorded. Then the explosive object can be disposed of according to existing procedures. To save the MED equipment in case of destruction in situ of a mine the special moving steel screen may be used for ballistic protection (see Figure 5).

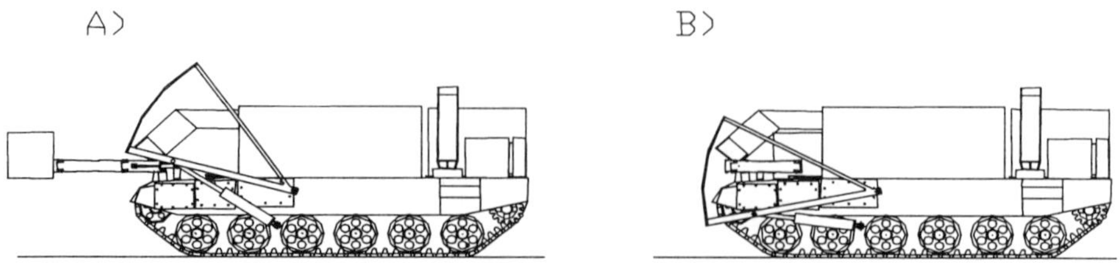

Figure 5. The MED with the steel screen. A) operating position, B) protected position.

Note that the tank chassis has some disadvantages: it is expensive, heavy, and its lifetime is limited. The example of the wheeled version of the MED shown at Figure 6 is cheaper and lighter and it has no problems operating on roads. In this case the mine detector and additional equipment are placed on a platform, which is moved by tractor.

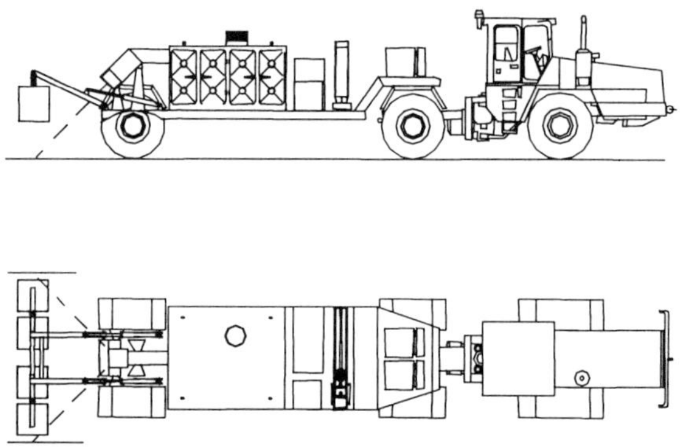

Figure 6. The MED on the wheeled vehicle.

5. The ecological safety

The landmine detection complex contains no radioactive sources and is safe when switched off. The local shielding around the target has mass 1500 kg. This is enough to reduce the radiation to a level that is safe for personnel at 33 m distance when the accelerator is switched on. The radiation level at distance 130 m is below the allowable radiation limits for general public. The residual activation of the soil, water and air is minimal since the time of the exposition is very short.

6. Conclusions

The discussed complex has the following advantages:

- It has high sensitivity and high probability of explosive detection. It satisfies the requirements of the UN clearance standards.
- It has high speed of inspection, which is 240-600 times higher than that for the hand clearance procedure.
- It has high selectivity and the expected level of false alarm signals is low.
- It can work on a terrain covered by bushes.
- It is able to inspect surfaces covered by water (beaches, rice fields, etc).

The additional advantage is that no equipment with extreme parameters is to be in the framework of the project. Prototypes of the main units of the complex (accelerator, particle detector and counting device, remote controlled vehicle etc.) have been already designed and tested for other applications.

Summarizing the above, the practical realization of the complex can help to solve the problem of humanitarian mine clearance.

References

International Standards for Humanitarian Mine Clearance Operations, (1996). UN-1996.-75p.

Trower W.P., (1993). The Nitrogen camera and the detection of concealed explosives, Nucl.Instrum&Meth., B79 (1993) 589

Belovintsev K.A, Karev A.I. and Kurakin V.G., (1987). The Lebedev Physical Institute Race-Track Microtron, Nucl. Instr. and Meth., B261 (1987) 36.

Belovintsev K.A, Bukin A.I., Gaskevich E.B., Karev A.I. and Kurakin V.G., (1994). The Radiation Complex for fundamental Research // in Proceeding of Fourth European Particle Accelerator Conferees. EPAC 94. London, 27-30 June 94, p.861.

Detection of anti-personnel landmines using microwave radiometry techniques

Markus Peichl, Stephan Dill, Helmut Süß
DLR (German Aerospace Center)
Institute of Radio Frequency Technology and Radar Systems
Oberpfaffenhofen, 82234 Weßling, Germany
Phone: +49-8153-28-2390, -2352, -2372, Fax: +49-8153-28-1135
E-mail: markus.peichl@dlr.de, stephan.dill@dlr.de, helmut.suess@dlr.de

INTRODUCTION

At present many million anti-personnel (AP) mines are polluting our environment in many countries. They cause a considerable limitation of the living space and make the land unuseable for agricultural purposes. Because the mine clearance procedures are much slower than the mine laying operations the number of polluting mines and the related contanimated areas are nowadays still increasing. Many mine polluted regions are former combat areas or belong to formerly highly frequented places. Thus the ground is contaminated additionally by many artificial impurities. For the current mine detection technologies and for the most modern mines, which become continuously smaller and have less metal content, a high false alarm rate of up to 1000 to 1 is therefore very likely.

Consequently, an advanced mine detection system has to fulfill more or less the following key requirements: It should be capable of detecting small surface and subsurface plastic mines with very low metal content under most environmental conditions; the detection probability should be maximised and the false alarm rate minimised; the system should offer a maximum in security and reliability for the operator; the hardware expense should be as most cost-effective as possible. Following our own investigations and the publications of many scientific groups of various working areas we conclude that the only way to approach the optimum solution is the use of multiple sensors based on different physical phenomena and principles and fusing their data. By doing this the different benefits of each sensor can be isolated from its shortcomings and merged to match as comprehensive as possible the specific mine detection requirements.

In this paper we illustrate the phenomenological background, the design, and experimental results of a multi-spectral low-frequency microwave radiometer as a part of a multi-sensor mine detection system. The overall system is intended to work in a hand-held operation allowing the use in areas of difficult access without excluding missions in more friendly environments. Thus the radiometer antenna is operated in an extreme near-field mode to achieve a corresponding ground resolution in the order of the active antenna aperture size. Additionally the radiometer receiver is swept in narrow-bandwidth steps through a broad frequency range to vary the penetration depth and the reflectivity properties of the actually observed ground part. This provides significantly increased information about the location and shape of buried objects for discrimination purposes.

PHENOMENOLOGY

Microwaves (MW) at longer wavelengths can penetrate dielectric materials in the order of centimeters of thickness depending on the actual frequency and the complex frequency dependent permittivity. The appropriate frequency range for subsurface mine detection in the order of

H. Schubert and A. Kuznetsov (eds.), Detection of Explosives and Landmines, 195–198.

centimeters is limited to the lower microwave region up to about 5-6GHz as shown in fig. 1. In fig. 2 the impact of the spatial resolution capability on the imaging of objects is shown. Demonstratively the receiving area of an antenna can be regarded as a flared cone leading to a distance dependent spatial resolution, which obviously becomes better for shorter distances. A hand-held microwave mine detection system can operate the measuring antenna in the near-field range where for the extreme case of a few centimeters distance the ground resolution is approximately given by the limit of the frequency dependent effective aperture size.

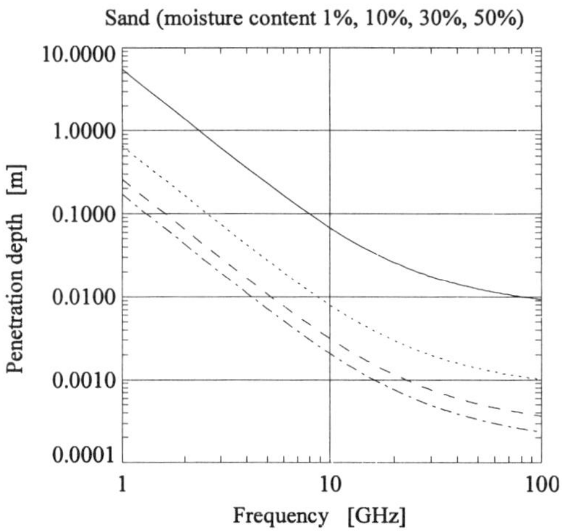

Fig. 1: Penetration depth for increasing moisture content from top to down.

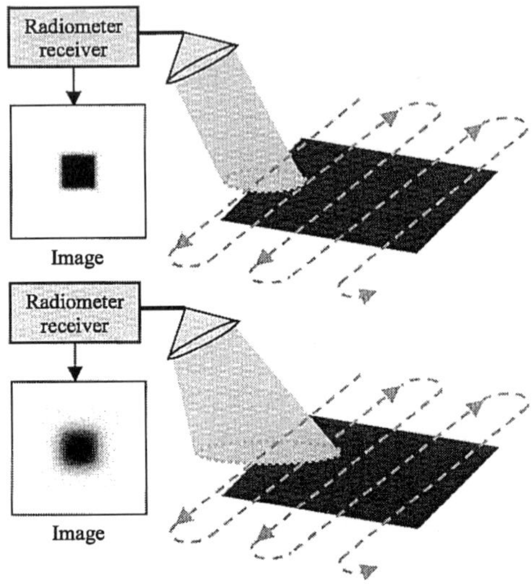

Fig. 2: Impact of the spatial resolution on the imaging of objects.

Basically, radiometers are very sensitive high amplifying frequency selective receivers. In contrast to radar they do not send any artificial signal but work in an receive-only mode and therefore cannot be used for range measurements in a similar way. The high sensitivity allows the measurement of variations of the naturally generated thermal radiation emitted by matter with a physical temperature above 0K. This radiation usually is composed of many contributions like the self emission of the observed object and the reflection and transmission of elsewhere similar generated radiation. For a top-down Earth observation as indicated in fig. 2 the reflectivity contrast of the observed object in comparison to the background is of major interest because of the radiometric cold, i.e. low power, behaviour of the sky radiation due to its cosmic origin [1, 2].

Furthermore, the structure of layered media, as for the case of buried mines, shows up in frequency dependent variations of the reflectivity due to interference and consequently in changes of the brightness temperature as a measure of the observed radiation power. The measured brightness temperature can change seriously with observation frequency as shown exemplarily in fig. 3 for two different moistured sand layers on a teflon layer. The graphs show the brightness temperature signals for various sand layer thickness whereby the horizontal curve indicates the pure sand signals without any teflon below. The frequencies and the amplitudes of the oscillations increase for higher frequencies and decrease for higher moisture content due to the impact of the complex permittivity and the according penetration depth. Consequently the adequate narrow-band sampling of a broad frequency band can supply important detection and discrimination information of objects when additional soil information is used. Therefore the investigation of the radiometric spectral near-field imaging of typical mine polluted scenarios in a frequency range of about 1-6GHz using narrow bandwidths promises to provide the most informative and useful results.

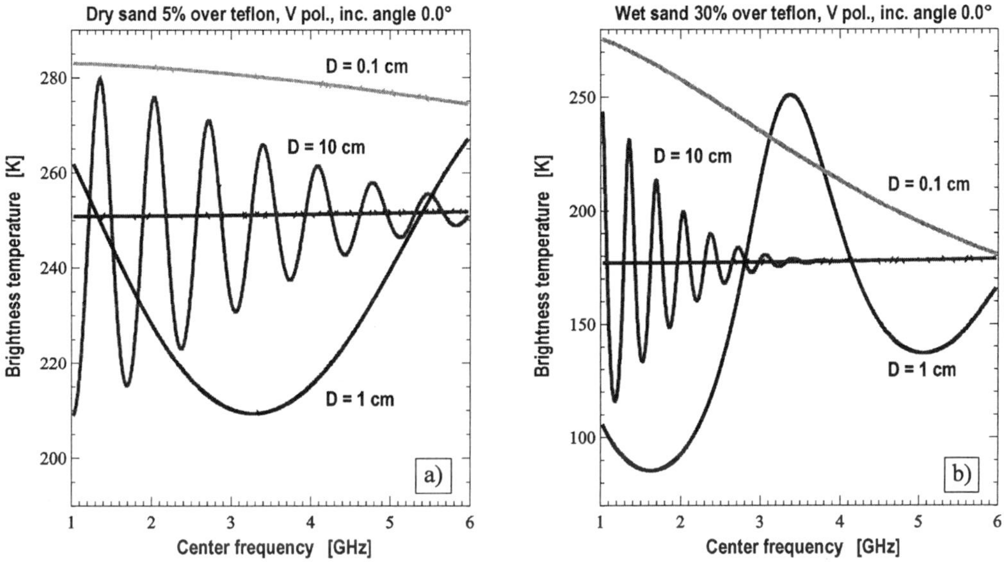

Fig. 3: Computed Nadir brightness temperatures of a layered structure air-sand-teflon in dependence of observation frequency and sand layer thickness. Figs. a) and b) show results for 5% and 30% of soil moisture. V and H polarisation are identical for Nadir observation.

EXPERIMENTAL SETUP AND RESULTS

For our first broadband spectral experiments a total power radiometer with a tunable filter of about 35MHz bandwidth was built up. Here a broadband Vivaldi antenna in a printed design is used because of its low size, weight, and costs. The overall receiver noise temperature is about 300K to 500K in the usable frequency range of 1.7GHz to 7GHz. An integration time of about 15ms for each center frequency can be used to achieve a desired temperature resolution of about 1K. We built up a two-dimensional computer controlled and step motor driven positioning system of 2m×3m size to perform raster scan images of typical buried mine scenarios. It is located outside together with two pits of the same size and 1m depth filled with sand and loosy soil and protected by a tent.

Fig. 4 shows the radiometer receiver, the ground Vivaldi antenna, and a sky Vivaldi antenna mounted together on one axis of the positioning system during a typical measurement arrangement. The sky antenna is used as a cold calibration source measuring the brightness temperature of the cosmic background radiation having about 3K. A 50Ω load of known physical temperature inside the receiver is used as a warm calibration source of about 320K. Both sources and the ground antenna are connected to a computer controlled switch integrated in the receiver. Fig. 5 shows a typical measurement result.

Fig. 4: Photograph of a typical raster scan measurement setup.

Fig. 5: Photograph, sketch, and measurement results of a buried mine detection scenario in sand for different center frequencies. The objects all were buried in different depths up to several centimeters.

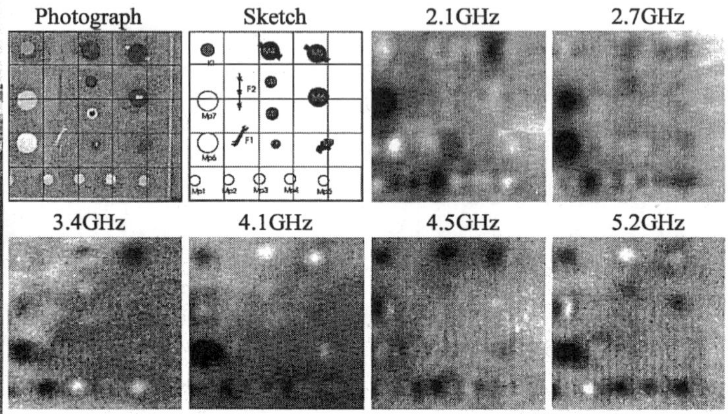

The mines and additional false and reference targets were buried in the sand pit in depths between 1cm and 10cm. The imaging was done by moving the broadband antenna in a raster scan very close to the surface across a desired area and sweeping through the frequencies. It can be recognised that the buried objects appear very different for the varying frequencies and show sometimes inversions in contrast due to the interference phenomenon. Furthermore, most of the low-metal or pure-plastic mines in the more or less right image parts can be detected. It showed up that the use of multiple frequencies allows an estimation of the depth and that the frequency signature and detected lateral object shape may be used as an indicator for identification and discrimination.

CONCLUSIONS

Promising results of a microwave radiometer for the purpose of landmine detection have been shown. This indicates a valuable contribution of this technique within a multisensor system to reduce the false-alarme rate, which is currently investigated within the HOPE (Handheld Operational Demining System) project initiated by the European Union. Here the combination of a metal detector, a MW radar, a MW radiometer, and a position monitoring system was selected [3].

REFERENCES

[1] F.T. Ulaby, R.K. Moore, and A.K. Fung, *Microwave Remote Sensing - Active and Passive - Microwave Remote Sensing Fundamentals and Radiometry*, Vol. I, Addison-Wesley, Reading, 1981.

[2] M. Peichl, H. Süß, S. Dill, M. Greiner, M. Zeiler, *Todays imaging technologies and applications of microwave radiometry*, Proc. of the German Radar Symposium GRS 2000, Berlin, Germany, 11-12 October 2000.

[3] HOPE Project Programme, ESPRIT 29870, EU, 1998.

ANALYSIS OF PERFORMANCE OF A SYSTEM FOR EXPLOSIVES DETECTION IN AIRLINE BAGGAGE.

VASILY V. SMIRNOV
V.G.Khlopin Radium Institute
28, 2nd Murinsky Av. St-Petersburg, Russia.

Preventing acts of terrorism directed to destruction of aircrafts by means of explosives hidden in passenger's baggage is a significant problem. It may be solved by thorough checking of airline baggage for presence of explosive materials and devices. One of the proposed numerous techniques for such checking, suitable for automatic lines of baggage control is based on detection of nitrogen which is a component common to most known high explosives [US Patent #3997787]. Nitrogen detection technique involves exposing baggage to thermal neutron flux and detecting characteristic gamma rays emitted by nitrogen nuclei in nuclear reaction:

$$^{14}N_7 + n_0 \rightarrow {}^{15*}N_7 \rightarrow {}^{15}N_7 + \gamma$$

Exited nuclei of $^{15*}N_7$ immediately decay to ground states emitting characteristic 10.8 MeV gamma rays. Probability or cross section of neutron capture by nitrogen nuclei is low, but high energy of gamma rays makes it possible to detect it against the background caused by gamma radiation emitted by other nuclei having higher probability of neutron capture. Nuclear data for nitrogen and a number of other nuclei which may lead to false alarms in nitrogen detection are presented in Table 1. These data show that number of gamma quanta emitted per one captured neutron is greater for nitrogen than for other nuclei in spite of it's low capture cross section for thermal neutrons.

Nucleis	Fraction in natural mixture of isotopes, %	Neutron capture cross section 10^{-28} m^2	Energy of gamma rays, eV	Number of qanta per 100 captured neutrons
Nitrogen-14	99.6	0.08	10.83	15
Magnesium-25	10.1	0.18	11.09	0.03
Iron-57	2.17	2.4	10.03	0.1
Boron-10	18.5	0.5	11.45	0.8
Silicon-29	4.7	0.3	10.6	0.2

Table 1

The most common and suitable detectors for energetic gamma rays are NaI(Tl) scintillation detectors having high efficiency of detection for gamma rays with energies around 10 MeV. Unfortunately energy resolution of these detectors is rather low and capture gamma radiation of some nuclei (Fe, Cr, Ni) may be partly detected in the "nitrogen window" (9.5-11.5 MeV). Pile up of pulses with small amplitudes may be too

H. Schubert and A. Kuznetsov (eds.), Detection of Explosives and Landmines, 199–201.

detected in this window at high count rates. Number of these pile up pulses detected in nitrogen window depends upon the count rate and the shape of the spectrum in the vicinity of this nitrogen window. Pile up leads to distortion of the shape of gamma spectra and this distortion cannot be fully corrected. To detect nitrogen by gamma-spectrometric technique means to extract its weak spectral component from strong background caused by capture gamma rays of other materials present in pieces of baggage in far greater quantities. Intensity and shape of this background may strongly vary with elemental composition of baggage content.

Experimental installation we used for detection of nitrogen containing explosives in airline baggage consisted of special chamber made of polyethylene with $4*10^7$ Cf-252 neutron source positioned in it's upper wall. Internal dimensions of the chamber were chosen in accordance with the mean dimensions of pieces of airline baggage. Gamma spectra were measured by four NaI(Tl) detectors 150 mm in diameter and 100 mm in height placed under the lower wall of the chamber. Positions of detectors and neutron source were optimized experimentally. Mean time of detection of 0.5 kg of TNT with 5% probability of false alarms was taken as a criteria of optimization. Geometry of the inner part of the chamber was optimized to increase thermal neutron flux and its uniformity across the chamber. Pieces of baggage were moved into the chamber on a special carriage. When the door of the camber opened the neutron source moved deep into the upper wall shielded by borated polyethylene and lead.

Block diagram of the experiment is shown on Fig.1

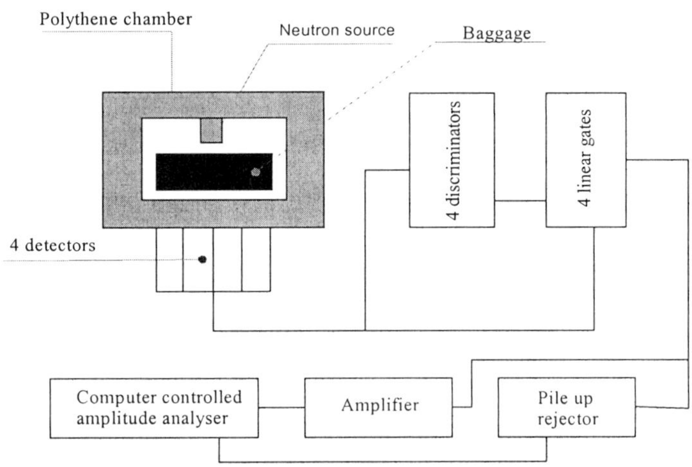

Fig. 1

Pulses from detectors were fed to single channel analyzers and linear gates opened by pulses from these analyzers. Pulses from all four linear gates with amplitudes exceeding 7.5 MeV were fed to common sum amplifier, pile up rejecter and computer controlled multichannel amplitude analyzer.

Number of pile up pulses in nitrogen window of each of the four separate spectrometric channels was about 10% of the net count rate in this window. When a piece of luggage containing large quantities of chlorine, chromium or nickel was placed in the chamber, number of pile up pulses could exceed number of pulses from 0.5 kg TNT nitrogen. Over 75% of these pulses were rejected. Stability of the system was controlled by the position of the 2.2 MeV peak corresponding to neutron captured by hydrogen atoms of polyethylene moderator. It was controlled for each detector separately with open linear gates. Count rate for 0.5 kg of TNT in 9.5-11.5 MeV region exceeded background count rates by 5-15% for different pieces of baggage.

Nitrogen content was determined by spectrum decomposition using least squares technique. Spectra of Cl, N, Ni, Cr, Fe and background were used as library reference spectra for decomposition of neutron capture gamma spectrum of each piece of baggage. Similar results were obtained when spectrum decomposition was performed by solving the system of 6 linear equations for 6 above mentioned components and number of counts in 6 regions of the gamma spectrum. Width and position of each region was optimized to minimum error in nitrogen content.

Duration of measurement needed for making decision about the absence or presence of nitrogen containing material in baggage with 95% level of confidence varied from 45 s to 15 min for different pieces. In the whole 150 pieces of baggage had been inspected, 70 of them contained 0.5 kg or 1 kg TNT. 8 false alarms and 4 failures had been observed. Errors occurred for objects with large quantities of chlorine, nickel and chromium and may be attributed to spectrum distortion caused by remaining pile up pulses.

Experience gained during the work shows that improvement in nitrogen detection characteristics may be achieved by using more effective rejection of pile up pulses and system stabilization. Number of errors may be significantly reduced when nitrogen distribution map across the inspected object is registered.

From report of T.Gozani [Gozani (1996)] it follows that they worked with systems of baggage control based on detection of nitrogen by 10.8 MeV neutron capture gamma rays. These systems had been tested in a number of airports. Over one million of pieces of baggage were checked and explosive materials were found in three of them. Experiments performed on these installations showed that sensitivity of reliable explosives detection is about 400 g of TNT which is in good agreement with our data.

References

Gozani T. (1996). Neutron techniques of nondestructive testing. Report on the conference on nuclear methods of detection of hidden explosives and fissionable materials. Obninsk, 08.04 – 11.04, 1996.

Sensorics of the German Mine Detection System MMSR

September 2001

C. Bittorf, J. Laukemper

Rheinmetall Landsysteme GmbH, Heinrich-Ehrhardt-Strasse 2, 29345 Unterluess Germany

KEYWORDS/ABSTRACT: metal detector, MSP, GPR, data fusion, real time processing

The future mobile mine detection and clearing system "Mobiles Minensuch- und Räumsystem" (MMSR) is designed to detect, mark and clear mines located on roads, paths and in open country. The MMSR program is now in the experimental phase, the aim being to test the effectiveness of the individual sensors in detecting mines, as well as their ability to operate jointly via data fusion. The following basic sensors have already been implemented: a multispectral camera platform, a magnetic induction sensor and ground penetrating radar. Each sensor has its own real-time signal processing, whose detection and classification results are merged in a higher order processing stage. Each time an alarm is generated a verification sensor checks for the presence of explosives in order to prevent false alarms. The mine-detection capability of the complete MMSR system will be tested in a specially laid minefield during the 4th quarter 2001, initially without the verification sensor.

1 Introduction

The mobile mine detection and clearing system MMSR is designed to enable combat engineers to detect, mark and clear mines on roads, paths and in open country.

By exploiting the latest developments in sensor technology, pattern recognition, image processing and robotics, the MMSR should be able to perform the following tasks:

- Assured detection (>98%) of exposed or concealed mines, whether metallic or non-metallic
- Marking of the mines and/or areas searched
- Neutralization/destruction of detected mines.

Figure 1: MMSR experimental carrier with integrated sensors: (EMI, GPR and MSP)

The MMSR is being jointly developed by a consortium of companies made up of:

- Rheinmetall Landsysteme
- STN Atlas Elektronik
- BGT Bodenseewerke

The MMSR program is currently in the experimental phase. The objective of this phase is to investigate the efficiency and effectiveness of the individual sensors in detecting mines, as well as

H. Schubert and A. Kuznetsov (eds.), Detection of Explosives and Landmines, 203–216.

their ability to operate jointly via data fusion. Based on the results of these tests, the requirements for the specialized sensors will be collated and used in preparing the concept for a future operational MMSR system. The current MMSR platform system (s. Figure 1) serves solely as a means of integrating the individual sensors, and has no relevance for a future mine clearing vehicle.

2 Overview

Figure 2 shows an overview of the basic subassemblies of the MMSR system. The sensors fall into two categories:

- Basic sensors
- Verification sensors

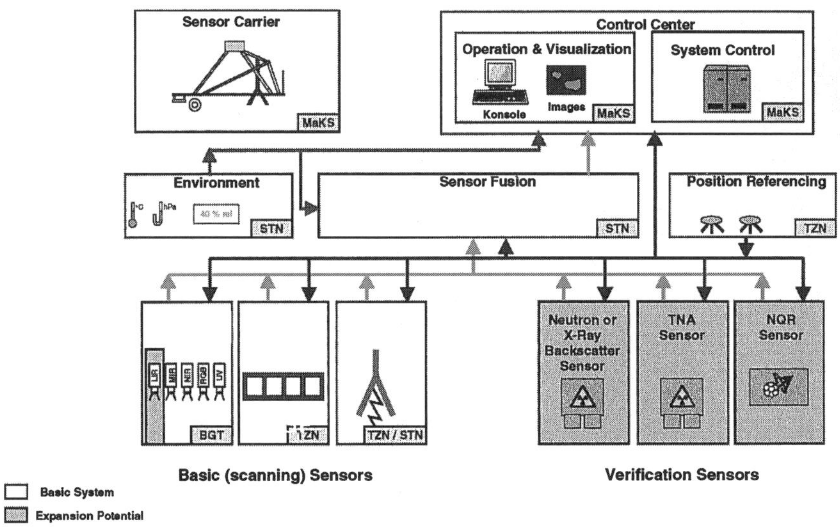

Figure 2: MMSR System Architecture

The basic sensors are all scanning systems (line or matrix), whose fields of view correspond to the search width of the experimental system (approx. 2 meters). Each sensor has its own signal processing unit, which evaluates and analyzes incoming raw data in real time. Each time a potential mine is detected, its precise location within the sensor's field of view and the accompanying feature vector are transmitted to a higher order data fusion unit. Each feature vector contains the sensor-specific characteristics of the object (e.g. metal content, size, and type of metal).

The basic sensors of the MMSR system are mounted in various positions on the sensor platforms, each having different fields of view, and collecting their data asynchronously to each other. In order to enable fusion of the individual sensor data, the coordinates of the detected object have to be transformed in a joint global coordinates system. To do this, the exact position and orientation (angular position) of the sensor platform are required during all measurement operations. This information is generated by means of a location reference sensor.

This sensor is based on a DGPS system operating in real-time kinematic (RTK) mode in combination with an inclinometer. Together with high-precision GPS time signals, the position and orientation of the sensor platform is recorded with a repetition rate of 5 Hz, and made available to all the sensors and the sensor fusion in real time. This produces a location accuracy in the centimeter region, and an angular accuracy of approx. 0.1°.

Fusion of the sensor data is carried out on the basis of the multi-sensor feature vectors and the accompanying position data of the location reference sensor. The fusion algorithm merges the information from the individual sensors via a fuzzy control unit, transmitting the fusion result to the control computer in the event of a mine alarm, where it is processed and visualized.

Each time the fusion process generates an alarm, a verification sensor (not currently integrated in the MMSR platform) checks for the presence of explosives at the spot where a mine is suspected. This leads to an appreciable reduction in the false alarm rate.

The technology for the verification sensor has still to be determined. The following techniques are currently being tested:

- TNA (Thermal Neutron Activation) analysis
- NQR (Nuclear Quadrupole Resonance)
- Neutron back-scattering
- X-ray back-scattering

3 Sensors

3.1 Multi-Spectral Camera Platform (MSP)

The optical sensor components of the MMSR system consist of a multispectral camera system for the UV, VIS, NIR and MIR spectrums (see Figure 3. The detection component features a lighting unit enabling mines to be detected regardless of the natural light conditions.

The optical sensors are mounted on a calibration-capable sensor platform approx. 3 meters above the ground. The optical axes of the sensors are aligned in parallel, perpendicular to the detection surface (i.e. the ground), so that mutual field of view of the optical sensors is maximized.

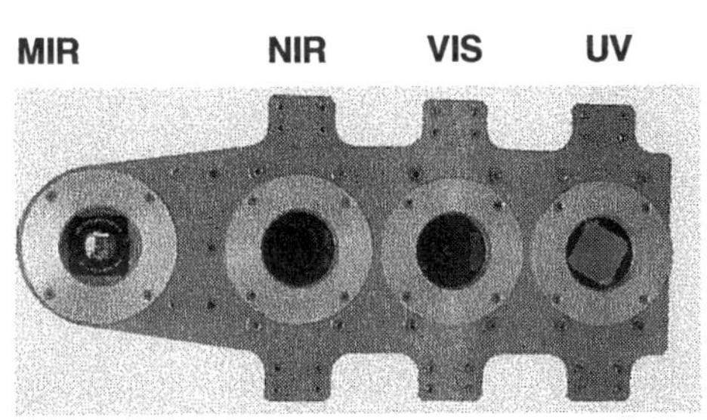

Figure 3: Multi Spectral Camera Platform

The optical detection components are characterized by the following performance data:

- spatial resolution on the ground < 1cm
- complete scanning of the ground at a maximum speed of 3 meters per second

The different wavebands are necessary to increase the detection probability and to discriminate between surface laid man made objects and the natural background.

As an example three images are shown in Figure 4 in the wavebands UV, NIR and IR. As one can see, the anti tank mine has a clear contrast in all bands, but the AP mine between the two plastic boxes can only be detected in the IR image. In other environments the situation can be vice versa.

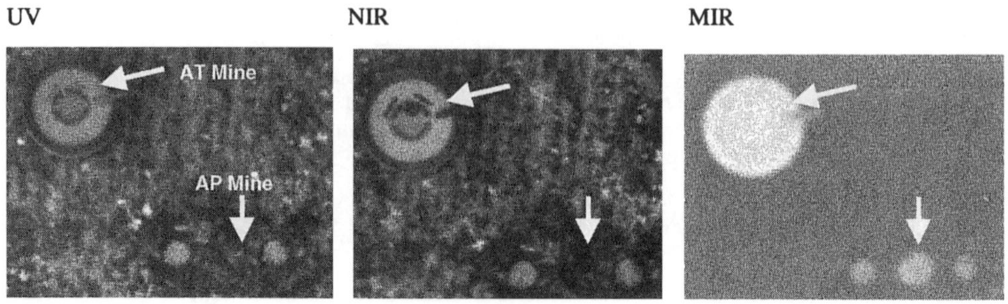

Figure 4: Images of surface laid mines in three different wavebands

Processing of the image information of all four cameras takes place on three different image processing levels, referred to as low-level, mid-level and high-level (see Figure 5), each characterized by different types of hardware architecture. Low-level processing encompasses image processing directly at the pixel level. The computer hardware used here is a systolic array processor specially developed for this application. The objective of processing in this case is to reduce the amount of data to the individual events represented by those pixels belonging to potential mines.

Each event is linked to a so-called event attribute containing the geometry and intensity characteristics of detected objects. Object segmentation and/or object labeling occurs at the mid-level processing stage on the basis of the individual object data (events) made available. For each image object, an accompanying set of event attribute features is extracted that characterize the corresponding object.

At the high-level processing stage, all pixel-related information from the extracted objects in each spectral channel is transformed in an inertial coordinates system, and then merged to an overall feature vector. Processing results in a list of potential mine candidates, accompanied by a list of characteristics, which, among other things, contain detection contributions from the individual channels.

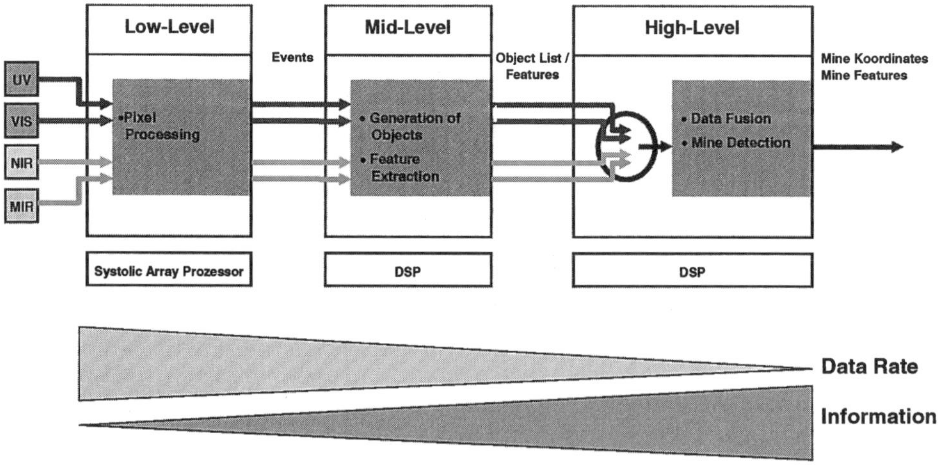

Figure 5: Signal Processing Architecture of MSP Detection Algorithm

3.2 Electro-Magnetic Impulse Sensor (EMI)

Within the group of MMSR sensors, the electromagnetic impulse sensor is responsible for detecting objects containing metal. The core component of the system is a multi-channel sensor array arranged in two levels one on top of the other. Figure 6 shows the metal detector in front of the MMSR carrier. The sensor is based upon the UXO detection system AMOS, developed by RLS but optimised for mine detection. That means small amount of metal of near surface objects can be detected with a high spatial resolution.

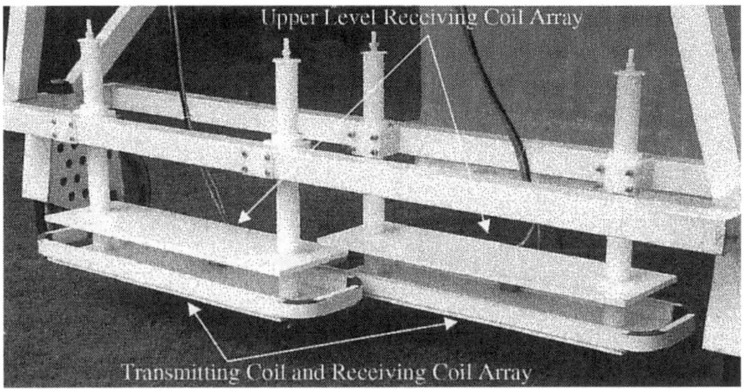

Figure 6: EMI Sensor within MMSR Carrier

The sensor is segmented in two parts for a better adaptation for a later operational sensor system. The sensor has two transmitting coils, which are electronically synchronised and every part has an array of 16 receiving coil in an overlapping configuration. The sensor covers a search with of approximate 2m.

Also the upper levels contain a linear receiver coil array of 16 receiving coil in an overlapping configuration. The arrangement of the receiver coils in two levels enables measurement of the gradient, thus guaranteeing exact determination of the depth of the detected objects. The electronics necessary for controlling the transmitter coil and for conditioning the receiver coil signals are integrated in a rack in the operator's cabin. The raw EMI data are transmitted to the signal-processing unit via a serial interface. Conducted in real time, the signal processing operation covers object detection, calculation of the object features, and transmission of the feature vectors to the higher order sensor fusion unit.

3.2.1 The Measurement Principle

The physical measurement principle of the electro-magnetic sensor is based on detection of an induced eddy current in a metallic body.

Active stimulation occurs via a pulsing transmitter coil, whose current is switched off for a time interval of several microseconds (impulse remanence method, see Figure 7). The temporal change of the primary magnetic field generated by the transmitter coil induces an eddy current in a metallic body located in the primary field. On account of finite conductivity, the eddy current in the volume and on the surface of the body diminishes exponentially with time.

This eddy current produces a temporally changeable magnetic dipole moment that correlates with the temporally diminishing secondary magnetic field. The strength and spatial distribution of the magnetic field is defined by the geometry of the body, its spatially distributed electric conductivity and permeability. The body (i.e. the object being searched for) thus represents the source of this secondary field, enabling the body to be detected.. Detection occurs by means of a coil array that serves as a receiver unit.

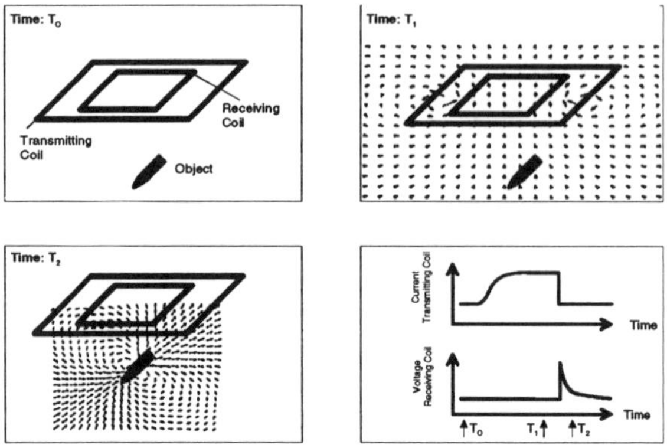

Figure 7: Diagrammatic representation of the impulse remanence method

The strength of the signal depends on:

- the strength of the primary field,
- the measurement time after switch-off of the primary field,
- the conductivity and permeability of the object material,
- the distance between the transmitter/receiver coil and the object
- the geometry of the object, and
- the spatial position of the object.

3.2.2 EMI performance data

The electromagnetic impulse sensor was optimized to detect even the smallest amounts of metal at shallow depths, while at the same time providing a high level of spatial resolution. The smallest detectable amount of metal is approx. 0.5 gram, with ferrous and non-ferrous metals being equally detectable. The spatial resolution is approx. 10 cm (see Figure 8), which means that two objects lying next to each other can be clearly resolved if the distance between them is not less than 10 cm.

The raw EMI data (see Figure 9)from the two receiver coil arrays are processed in real time and visualized in a false-color image. Detection of objects and subsequent feature extraction is carried out autonomously by the signal-processing unit. Subpixel processing permits objects to be located with an accuracy of < 5 cm. Apart from the exact location, the extracted features also indicate the size (diameter), the depth, and the type of material (ferrous or non-ferrous) of the detected object. To each feature a measure of confidence is assigned, which describes the confidence coefficient of the respective reading.

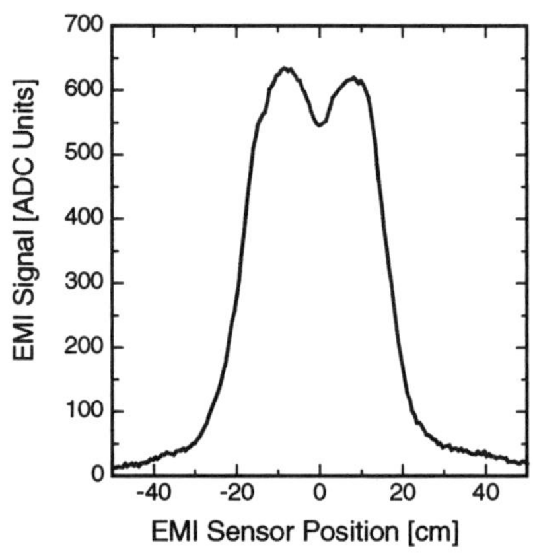

Figure 8: EMI signal during scanning of two iron balls located 15 cm from each other.

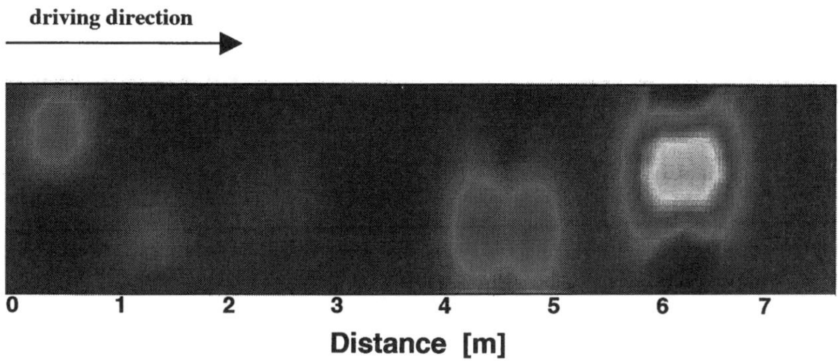

Figure 9: Raw EMI data visualization of some detected objects with different amount of metal

3.3 Ground Penetrating Radar (GPR)

The task of the ground penetrating radar (GPR) sensor within the MMSR group of sensors is to detect non-metallic and metallic antitank and antipersonnel mines, whether exposed or concealed. The sensor is designed for a measurement distance of approx. 25 cm above the ground.

Specially developed for this application, the GPR system is based on an antenna characterized by a high signal-to-clutter ratio, which avoids multiple reflections of the electromagnetic waves in the antenna inside the observation timeframe.

3.3.1 GPR Principle of Operation and Limitations

The GPR used is a pulse/ echo-type system. Short broadband radar impulses are directed into the ground, and partly reflected by every dielectric inhomogeneity. A receiver antenna receives the reflected radiation during the preprogrammed time gate.

The essential limitations entailed in using this process are:

- the dispersal attenuation
- the antenna clutter

The dispersal attenuation of electromagnetic waves in the ground depends on the type of ground and the ground moisture.

It is generally true that high frequencies permit a high degree of spatial resolution, but due to strong attenuation in the ground they cannot penetrate the soil very deeply. In accordance with the task at hand, a compromise choice must be made with respect to pulse form and signal bandwidth.

Another basic problem with the GPR system relates to the radar antenna system itself. A radar antenna represents a source of disruption for electromagnetic waves. This results in multiple reflections, e.g. between the ground and the antenna, and also in multiple reflections within the antenna caused by the antenna mounting, connectors, etc. These reflections are detected by the receiver elements, leading to the undesirable extra signals known as antenna clutter. They interfere with the often very weak signals coming back from the target, which are strongly attenuated in the ground. If the antenna clutter is too strong, the target signals can be blocked out completely.

Antenna clutter can be minimized only when the radar antenna features the lowest possible radar -scatter cross-section.

3.3.2 GPR System Design

In the case of an antipersonnel mine with an assumed thickness of $a \approx 4\text{cm}$ and a dielectric constant of $\varepsilon_r \approx 3$ (explosive), a time of flight difference for the radar reflection from lid to ground of $\Delta t \approx 0{,}46\text{ns}$ results. In order to resolve these echoes, the radar pulse width T has to be $T < \Delta t$. This results in a radar signal bandwidth of at least 2.5 GHz.

On the other hand, due to the high ground attenuation, frequencies of > 3 GHz should be avoided. These requirements point to a radar pulse form consisting of only one oscillation (see Figure 10).

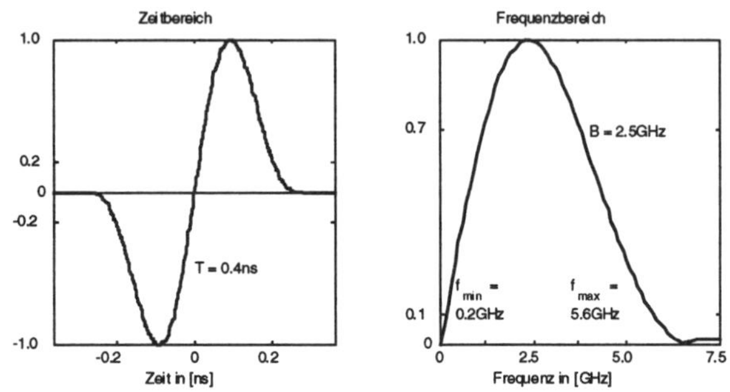

Figure 10: Broad band Radar Pulse (time and frequency domain)

Pulse forms of this kind display an extremely large maximum to minimum frequency ratio. A system for processing signals like these must have extremely broadband characteristics (ultra wide band (UWB) systems). A traveling wave antenna for example offers this broadband characteristic.

The simplest form is a dissipative (resistively loaded) dipole, in which the conduction current has been largely absorbed by the time it reaches the end of the dipole due to line losses, while the electromagnetic waves de-couple themselves along the dipole into free space. Additionally, using a V-shaped dipole arrangement, a directional effect can be achieved. In order to meet the requirement for an antenna that is as clutter-free as possible, the entire antenna must be free from any internal reflections. This was achieved by means of a special antenna design. (see Figure 11).

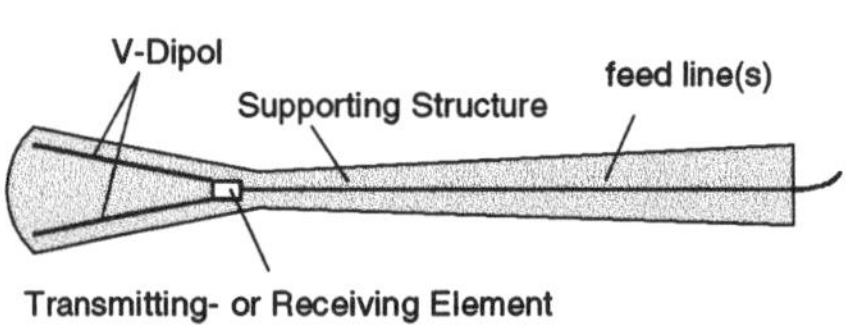

Figure 11: Cross section of GPR antenna segment

The transmitter and receiver elements are directly integrated in the base of the V-dipole. Thanks to the application of SMD and hybrid technologies, the dimensions of the transmitter and receiver elements are distinctly smaller than the radar wavelengths, thus avoiding reflections. The power supply cables are laid in a 1.7-meter-straight line to the antenna elements in order to keep this area of the antenna free from reflections as well. The length of the lead was selected so that the time of the waves running from the base of the antenna along the power cable and back is longer than the observation time of 10 ns.

Because the antenna components (V-dipole, transmitter and receiver elements, power cable) are not self-supporting, they need a carrier structure. It too has to fulfil the requirement of interfering with the propagation of radar waves as little as possible. The material used for the carrier is Styrodur, which features a dielectric constant of 1.2, and thus comes very close to meeting the requirement. To protect it against the weather and improve its mechanical stability, the entire antenna is encased in Macrolon.

In order to achieve a detection width of approx. 2 meters, the antenna consists of a linear array of twenty transmitter and twenty receiver elements, each situated five cm apart in an alternating row (see Figure 13).

Groups consisting of two transmitters and two receivers each are controlled by one multiplexer. This enables a directional characteristic perpendicular to the dipole to be achieved, resulting in a 15-cm footprint at a distance to the ground of 15 cm.

The observation interval of 10 ns is depicted via a sample receiver in a timeframe of 1.28 ms, and scanned

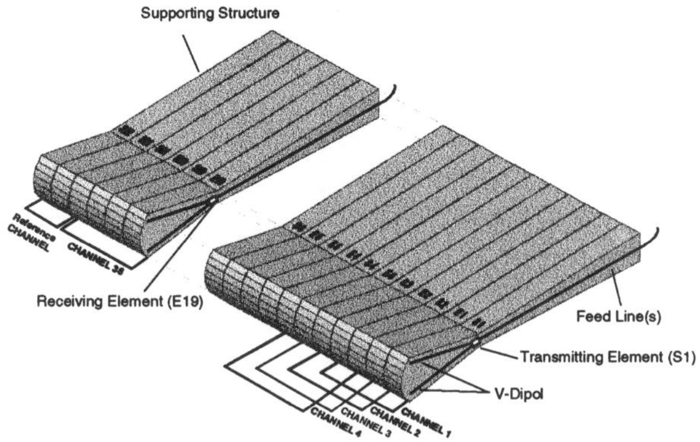

Figure 13: GPR Antenna Assembly

Figure 12: GPR Antenna integrated in MMSR Ca

with 512 points. Thus, a complete GPR image consists of 38 x 512 (width x depth) pixels and is recorded at a data rate of 20 Hz.

Figure 14 through Figure 16 show raw GPR data from various mines in the form of grayscale pictures. During measurement, the antenna was moved over the object at a speed of approx. 0.15 m/s. The abscissa of the images corresponds to the sample values in the movement direction. The grayscale reproduces the level of the signal amplitude of the reflected radar pulse. On account of the sinusoidal shape of the transmitted radar pulse, each reflecting boundary layer (step in the dielectric constants) shows a light /dark or dark/light transition, depending on the type of dielectric transition.

The upper partial picture shows a longitudinal cross-section of an individual antenna element in the movement direction. The ordinate corresponds to 512 sample values within the observation window, and thus to the depth values in the ground.

The lower partial picture represents a horizontal cross-section within the observation window with a fixed sample value, i.e. a cross-section at a fixed depth. The ordinate corresponds to the 38 antenna elements.

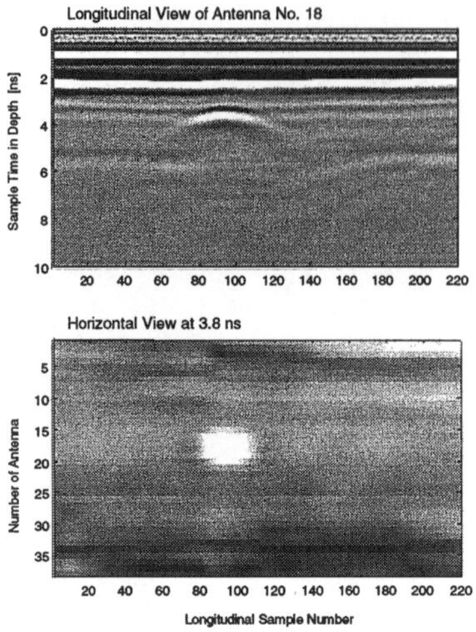

Figure 14: GPR Raw Data Images of an Anti-Personnel Mine (Plastic case with metal strap)

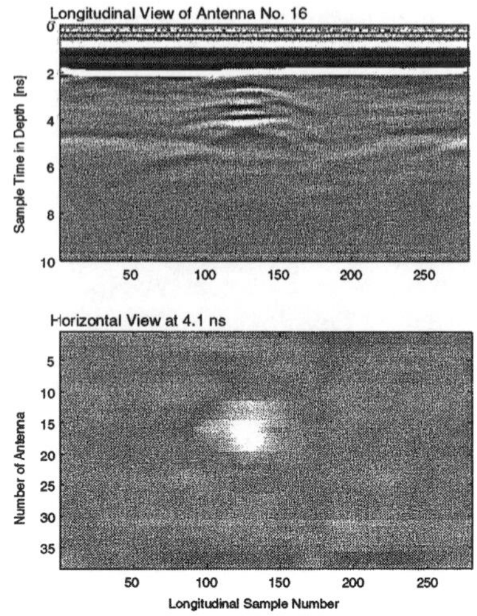

Figure 15: GPR Raw Data Images of an Anti-Tank Mine (Plastic case; metal only in the mine fuse)

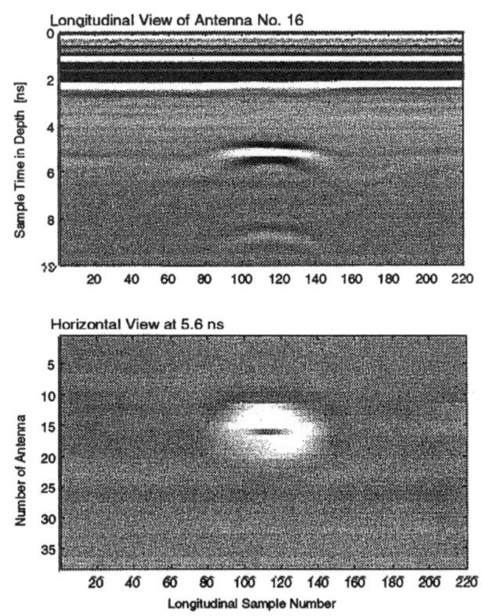

Figure 16: GPR Raw Data Images of an Anti-Tank Mine (Metal case)

3.4 Explosives Detector

3.4.1 Operational Concept

The sensors described in the preceding chapters,

- the multi-spectral camera platform (MSP)
- the electromagnetic impulse sensor (EMI)
- and the ground penetrating radar (GPR)

are the basic sensors of the MMSR system. The data fusion unit makes its decision on a possible mine alarm based on the detection and classification results produced by these sensors. Each alarm is characterized by a classification probability, the precise position of the object, and a feature vector (see Figure

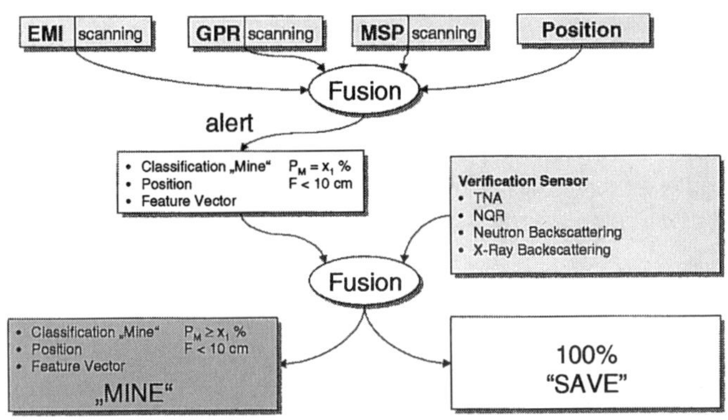

Figure 17: Detection flow chart of MMSR System

17). When an alarm is triggered, the verification sensor is activated. It is positioned over the location of the object detected by the fusion unit, and checks the adjacent area for the presence or non-presence of explosives, with the main focus being on the last-mentioned point.

The main task of the verification sensor is to reduce the number of false alarms, not to enhance the classification probability of a supposed mine alarm.

3.4.2 Possible Technologies for Verification Sensors

The verification sensor has to be able to detect bulk explosive with a high degree of certainty. The following technologies largely meet this requirement:

- TNA (Thermal Neutron Activation) Analysis
- NQR (Nuclear Quadrupole Resonance)
- Neutron back-scattering (restricted)

A further alternative is offered by x-ray back-scattering technology. Although it does not fulfil the requirement of being able to detect explosives with a high degree of certainty, this method can generate a spatial three-dimensional image of the interior of the object structure. Thus, under certain circumstances, this technique could permit identification of buried mines, meaning that it too could be used as a verification sensor.

Thermal Neutron Activation

TNA technology uses thermalized neutrons, which can be generated either in a nuclear isotope source or a neutron generator, to stimulate the nitrogen nuclei of the explosive. The stimulated nuclei emit energy in the form of characteristic gamma rays, which can be used to prove the presence of explosives.

The TNA technique features the following advantages and disadvantages:

Advantages of TNA technology:

- high penetrating power of the neutrons
- the amount of explosive material can be determined

Disadvantages of TNA technology:

- the presence of explosives is proved only indirectly via the nitrogen nucleus (cross sensitivity)
- ground moisture reduces depth of penetration
- the type of explosive cannot be determined
- heavy shielding measures are necessary between the detector and the neutron source

Nuclear Quadrupole Resonance

The nuclear quadrupole resonance technique is a highly sensitive spectroscopic method of examination in which resonance frequency of a nuclear spin transition is measured by irradiation with radio frequency pulses. NQR exploits the electromagnetic interaction between atomic nuclei having a permanent quadrupole moment (for example nitrogen) and the internal electric field gradient due to the molecular structure of the substance.

The NQR technique features the following advantages and disadvantages:

Advantages of NQR technology:

- simple, compact front-end is possible
- low cross-sensitivity vis-à-vis other substances
- type and amount of explosive can be determined

Disadvantages of NQR technology:

- TNT can be identified only with great difficulty
- metal-encapsulated explosives cannot be detected

Neutron Back-scattering

Neutron back-scattering technology is not a means of explosives detection as such, since explosives are detected only indirectly via their hydrogen nuclei. In contrast to TNA analysis, with the neutron back-scattering method fast neutrons are employed as a means of inspection. The hydrogen nuclei of the explosive slow down the fast neutrons through an impact process (thermalization). The detection of scattered, decelerated neutrons indicates the presence of light hydrogen atoms, and thus, indirectly, of explosives.

Neutron back-scattering technology features the following advantages and disadvantages:

Advantages of neutron back-scattering technology:

- high penetrating power of the neutrons
- a simple, compact front-end is possible

Disadvantages of neutron back-scattering technology:

- extremely susceptible to ground moisture and vegetation
- the presence of explosives is proved only indirectly via the hydrogen nuclei

X-ray back-scattering

With the x-ray back-scattering method, a x-ray beam is used for the inspection rather than neutrons. By measuring the intensity of the reflected x-rays, information becomes available on the density of the material. By means of a scanning device in connection with a collimated x-ray source and a detector spatially limited by apertures, a spatial image of the object structure is generated, which can be used in identifying the object.

Back-scattering technology features the following advantages and disadvantages:

Advantages of x-ray back-scattering technology

- 3D image-generation

Disadvantages of x-ray back-scattering technology

- limited ground-penetrating capability
- very long measuring time
- elaborate, heavy shielding is necessary

4 Present State of Work and Future Plans

As it now stands, the MMSR experimental system has been equipped with a complete location reference system and the basic sensors, e.g. the multispectral camera platform, magnetic induction sensor and ground penetrating radar, including the necessary electronics and signal-processing equipment. Development of real-time signal-processing for the camera platform, the metal detector and the GPR is now complete.

At present, the data fusion contains the control unit for linking of all sensors. An extension of the fusion unit with respect to the verification sensor can be easily done, because the whole system architecture is an open architecture. Subsequently, the complete system will be tested and system performance will be investigated until the 4th Quarter of 2001.

In order to test the MMSR sensor systems, several 50-meter-long minefields were laid out, each featuring different vegetation and ground structures. Each minefield contains between 10 and 15 mines (AT and AP), laid in different ways. Most of the mines contain original explosives (the detonating canal was interrupted) as a prerequisite for testing the MMSR in a subsequent phase in conjunction with a verification sensor (explosives sensor).

The technology selection for the verification sensor is not yet completed, because some new and promising technologies are appeared in a very early state of development. Our proposal is the technique of the prompt neutron capture gamma analysis (PNCG).

The neutrons from a neutron source (Cf 252) were moderated by the moisture of the soil and/or the hydrogen content of the explosive. The now thermal neutrons can be captured, our interest is the capture from nitrogen as an indicator for explosive material. With a relative measurement technique (reference point beside the mine / suspected area) is it possible to eliminate the background (artificial fertilizer).

The main assemblies of the front end of an PNCG Sensor are(see Figure 18):

- Two sources Cf 252 with a special shutter system, if the front end is not in operational mode
- BGO detectors for the measurement of the gamma radiation
- Neutron shielding for the BGO detectors
- Gamma shielding against the neutron sources
- A moderator to decrease the neutron flux in all other directions than to the ground
- Biological shielding for the protection of the personal

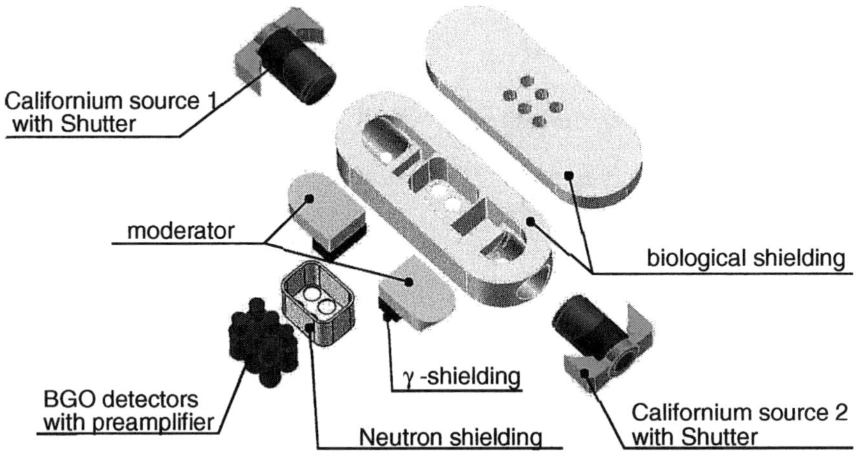

Figure 18: Main assembly of the future PNCG sensor for the detection of explosives

The weight of the future front-end will be 180 kg. Some technical data for the PNCG Sensor:

Neutron source:

Isotope	Californium 252
Quantity	0,1 mg for both sources
Neuron flux	$2 \cdot 10^8$ n/s
Specific activity	2 GBq or 53,6 mCi

System Parameters:

Measuring time	< 60s
Maximum of depth penetration	30 cm (depends on the humidity of the soil)
Minimum detectable explosive mass	app. 100g

THE PROBLEM OF MILITARY TNT IN NQR MINE DETECTOR

V.S.GRECHISHKIN
Kaliningrad State University,236041 Kaliningrad,St.A.Nevsky 14

1. Introduction

1.1. HISTORY OF NQR RESEARCH AT KALININGRAD STATE UNIVERSITY

One of the most promising methods of demining is nuclear-quadrupole resonance (NQR) with nitrogen-14, which is contained in practically all mine explosives. In NQR the quadrupole moment of a nucleus interacts with the electric-field gradient of the surrounding molecule and crystal, and thus the energy-level differences are molecule- and crystal-specific. In pure NQR, no external magnetic field is needed. One may, however, combine nuclear magnetic resonance (NMR), e.g., of protons, with NQR to achieve selective excitation or better signal; in this case, a strong magnetic field has to be switched on and off.

NQR is very important for humanitarian demining, because this method was shown to have very high success rate (97%) in field experiments in soil. The pioneer work in applying the NQR technique to explosive' detection in Russia was done by my group at the Kaliningrad State University.

Our work on applying local NQR to explosive' detection was begun by Prof. V.S.Grechishkin, Dr. S.I.Guschin and Dr. E.M.Shishkin in 1972. The experiments were carried out with samples containing Cu_2O (NQR of nuclei $^{63,65}Cu$) and metallic cobalt (NMR of nuclei ^{59}Co). The best results were achieved on observation of NMR of ^{59}Co, with the distance between the transmitting antenna and the sample being 15 m; the distance between the reciever coil and the sample was 0.4 m. Experience obtained in these measurements [Grechishkin (1973)] was later applied to other chemical compounds.

Starting from 1978 at Kaliningrad we have studed NQR and EPR in more than 2000 complexes. In three years a spectrometer operating in 1-8 MHz frequency of local NQR was created. This construction is different from other spectrometer [Azisov (1978)], which contained the scheme of double frequency transformation. The transmitter of our spectrometer was generating the pulse series as spin-locking solid echo (SLSE), two pulse Hahn spin-echo, strong off-resonance sequence (RAM) [Anferov (1990)]. The pick power of transmitted radio frequency pulse was up to 1 kW. In the process of registration of NQR signals were using synchron detector and digital accumulator AI-4096.

Afterwards there were researches of different methods detecting weak NQR signals which included SLSE, RAM, MW-4, MW-2 and WHH-4 [Grechishkin (1993,1997)]. The flat surface coil was used as transmitter-receiver coil. The main conclusion from these

H. Schubert and A. Kuznetsov (eds.), Detection of Explosives and Landmines, 217–225.

studies was that the useful sequence for distance detection of NQR was RAM. In 1984 we carried out local measurements of explosives as RDX, tetryl and HMX. The first measerements were carried out using the first version of multipulse spectrometer developed by F.Azizov. We have used the Carr series for detection of explosives (PVV-4, Semtex) for this spectrometer. Later, we observed NQR signals from Soviet mine TM2-62P, Italian mines TC-2.5,TC-6 on new spectrometer. The mine detection was based on searh of RDX NQR signal. At that time it was possible to detect one mine in a square of 1.5m×1.5m in soil in 5 minutes. The depth of mines in soil was less than 13 -15 cm. Unfortunately, it is not possible to detect military TNT in this conditions. Nevertheless, during 1978 we investigated different mixtures containing TNT by a double NQR-NMR method. For the anylysis of the chemical mixture containing different compounds of nitrogen-14 the two frequency method of identification was proposed in pulsed NQR [Grechishkin (1993,1997)].

Later, our group designed a new spectrometer for local NQR studies. The main innovation in this spectrometer was application of ferrite receiver antenna with time sharing. The device allowed us to achieve the detection distance 25 - 35 cm for the detection of RDX, HMX, TNT and tetryl.

In 1997-1998 we have carried out NQR tests in Perth, Australia. We were using analog and digital receivers with quadrature detectors for Fourier transformation. The comparison of Fourier, Hilbert and Hartley transformation was made. We used Hewlett-Packard mainframe E1401A with two ADC HP E1437A, signal processor HP E1485,VX 1 BUS HP 1482 B and Hewlett-Packard workstation B132L. Signal from the respective differential amplifier was conducted to the ADC. Unfortunately we were not able to improve the previous characteristics, because FFT and composite pulses are not working at low signal-to-noise ratio.

1.2. MAIN CHALLENGES IN DETECTING TNT BY THE NQR TECHNIQUE

Below is the list of the main problems which must be overcome when the NQR is used as a mine detector:

1. The influence the isomorphous impurities in military TNT and military mixtures of explosives in mines. 3% of isomorphous impurity reduces the intensity of NQR signal in 2,4,6-TNT 10 times.
2. One can use such impurities, especially in mines, to deliberately mask them from NQR-mine detector.
3. The complexes of 2,4,6-TNT with organic compounds give broad NQR lines in such mixtures.
4. The influence of EPR signals in charge transfer complexes at room temperature.

For RDX, HMX, tetryl, urotropin there are no problems in remote pulsed NQR. But for TNT it is better to use remote DNQR at the distance 20 cm with half-toroidal coil. Of all TNT types only monoclinic TNT present some difficulties. For purification of military Russian TNT we have used method of melting zone and received monocrystal of 2,4,6-TNT. Later we studed NQR in military TNT at room temperature. In this works we were using mechanical field cycling in DNQR and detected as impurity p-nitrotoluene at frequencies 260, 940, 1200 kHz at room temperature (T_{1d}=0.8 s, t_{pol}=360 s). But in the last measurements in pure p-nitrotoluene the NQR frequencies in electronic field cycling were 230, 884, 1114 kHz at room temperature (T_{1d}=0.8s, T1p=50s, t_{pol}=160s). TNT is a very good acceptor and it was possible to observe charge transfer complex 1:1 with TNT.

To overcome the small TNT pure-NQR signal, two methods are to be looked at:

- Detect TNT by direct NQR of its impurities: These impurities are usually present in per cent ratios (e.g., di-nitro-toluene, hexa-nitro-stilbene). If their NQR signals are markedly larger than the one of TNT proper, detection could rely on them.
- Detect TNT by double-resonance/cross polarisation with protons: By applying a strong magnetic field, proton NMR levels differ strongly in energy and are selectively populated. When the field is reduced to (near) zero and the energy difference equals that of nitrogen NQR, the proton polarisation can be transferred to neighbouring nitrogen nuclei. This can be detected directly at the nitrogen NQR frequencies, or - after switching on the magnetic field again - indirectly at the proton frequency.

The applicability of these methods have been demonstrated in the laboratory. For field application, RF excitation and detection has to be done by a surface coil with the sample (the mine) at a few decimetres in front of the coil. Surface coils for RF have been built and demonstrated by the NQR group in Kaliningrad. In method 2, the strong magnetic field has to be produced by a surface coil as well. This has not yet been done and presents a formidable task and needs an experienced post-doctoral scientist.

With method 2, there is the problem that some anti-tank mines may trigger on a strong magnetic field. One will have to analyse the scale of this problem (use of such mines in Combogja) and propose solutions; these could involve e.g. first testing with a dispensable permanent magnet on a moveable arm in front of a safe vehicle.

2. Design and characteristics of the TNT detection system based on NQR technique

As whole, an NQR landmine detector is a subject to three design criteria. Firstly, the detector must have high sensitivity and accuracy in detection of signals with low signal-to-noise ratio (SNR) of the separate NQR multiplets from the compounds to be detected in remote NQR experiment. Secondly, there must be a possibility to detect simultaneously several compounds, such as TNT, hexogen, and tetryl. Thirdly, the phase stability of the system must be sustained for the work with broad NQR lines, whose spectral width is up to 100 kHz. The main difficulty in detection of TNT in mines, compared to other compounds, is that in the process of manufacturing land mines after priming of a molten TNT and its cooling off inside of mine's plastic case, the resulting becomes a solid solution with NQR spectral width up to 100 kHz, what is much more difficult to detect than compounds having narrow, well separated lines.

The developed system is an improvement of the system reported in [Grechishkin (1993, 1997)] and is specially designed for the simultaneous detection of two lines of TNT at 740 and 840 kHz, those of tetryl at 5290 kHz, and of hexogen at 5192 kHz. The block-scheme of the system for TNT is shown on Figure 1. Quartz-crystal oscillator synchronizes the pulse generator, which in its turn handles the gates 1 and 2, and the oscillators 1 and 2 of the next stage, producing the RF oscillations at 740 and 840 kHz correspondingly. After that, the signals from the oscillators 1 and 2, having sine-like waveshapes, are transformed by the wave-shapers (WS) 1 and 2 of the following stage, into those of the meandr pattern, which is preferable due to its high phase stability. Then signals are logically summed in the logical unit "either", and then are transmitted into power amplifier, which amplifies the initial RF pulses to the 0.5 kW power each. Then, the pulses with two carrying frequencies that are the NQR frequencies of TNT are applied directly to the coil system. The coil system consists of a one-loop irradiating flat coil and four ferrite receiving aerials. The ringing in the receiving coils during the RF pulses is damped by the Q-switch circuit coupled to the coil system. After amplification of the received NQR signal, it was mixed in two channel phase sensitive detector of the Shuster type with the SYNC signals from both oscillators 1 and 2. As the expected maximum spectral width of the signal from TNT is about 100 kHz, the sampling rate in ADC is set equal to at least 5 μs for the accuracy of the convertion.

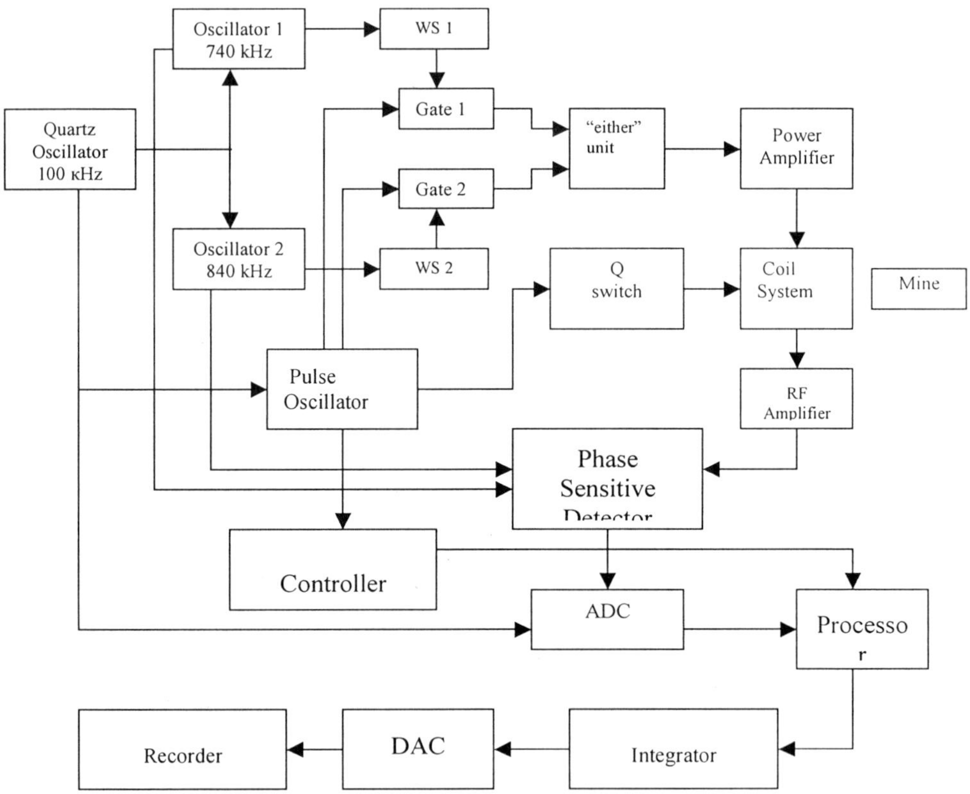

Figure 1. Block scheme of the NQR mine detector.

Signals obtained during tuning of the receiving ferrite coils, also coupled inductively with the irradiating coil for detection the signals of TNT and those of hexogen and tetryl are shown in Figure 2 and Figure 3 correspondingly.

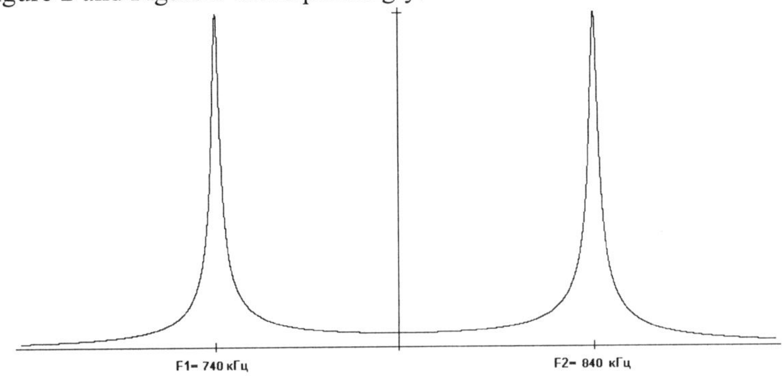

Figure 2 Tuning of the receiving ferrite coils for the detection of TNT.

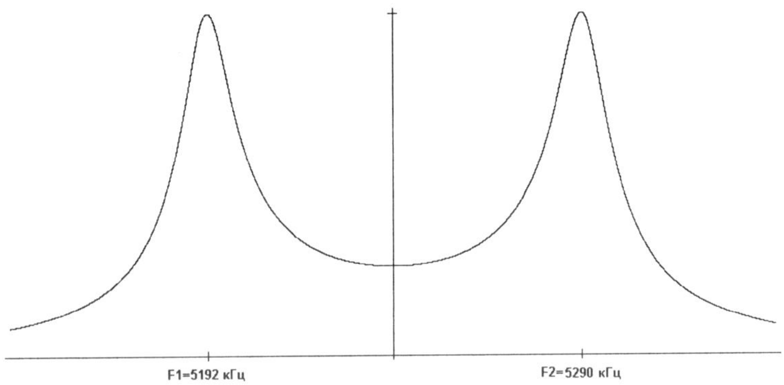

Figure 3 Tuning of the receiving ferrite coils for detection of hexogen and tetryl.

Besides the solutions of the explosives, the widespread solid solution of paranitrotoluene and TNT had been investigated. An addition of the 7% of TNT to paranitrotoluene leads to broading of its spectra about ten times at the NQR frequency 1144 kHz, and so for its detection use of stimulated echo sequence is necessary, while for detection of the mines TC-2.5 and TC-6 the SORC sequence is applied. The latter sequence was also applied for detection of M-14 mines containing tetryl. Partial Q-factor of the tuning curve in Figure 2 is up to 460, and with use of the Q-amplifier it is up to 4600, which allowed us to detect a TC-2.5 mine at distance 8 cm in two frequency mode. The tuning has been accomplished by tuning capacitors and changing the coupling between irradiating and ferrite coils, which in its turn lead to a change of the partial Q-factors whose working values are (in normal circumstancies) about 450 - 500.

In our work on applying the NQR technique to the detection of TNT by direct NQR of its impurities we have pursued the following tasks:

- search for information on TNT impurities, crystal structure etc. for different

countries, producers, mine types
- design, build/assemble 0.5-5 MHz probe head and electronics
- measure NQR properties of the relevant impurities in laboratory using traditional solenoidal coil with the sample inside
- estimate signal, minimum mass for detection at e.g. 0.3 m distance using surface coil (sample outside of coil)
- design and build demonstration system with RF surface coil(s) - optimise field distribution, materials, cooling, quality factor, electronics,high temperature supercoducters
- demonstrate measurement of TNT using its impurities, or model substance (e.g. sulfanylamide at 1-2 MHz).

In order to design and build surface coil for strong magnetic field for double resonance/cross polarisation, the following aspects were studied:
- model field distribution for variable geometry (thickness and width of windings, stacked flat coils, three-dimensional arrangements) and optimise field distribution
- analyse systems aspects (conducting material, power supply, current switching, cooling, integration of RF coil(s), shielding, etc.)
- study problem of depth resolution by scanning the proton frequency using the inhomogeneity of the magnetic field
- design and build optimal integrated surface-coil system
- demonstrate measurement of TNT or model substance using cross polarisation at various distances in front of the surface coil.

The detecting system [Anferov (1990)] was tested only in the laboratory conditions, while other more reliable and conveniently-tuned, system [Azisov (1978)] underwent multiple testing in the field conditions in Borisovo, Novosibirsk and Nakhabino apart from its preliminary laboratory test. The system for detection of TNT ,tetryl and hexogen was tuned at NQR frequencies 5290 and 5192 kHz correspondingly, which allows one to detect both compounds simultaneously; this feature can be used, for instance, in looking for the Italian plastic case mines TC-2.5 and TC-6 having 45% of hexogen and 55% of TNT. Such system can detect hexogen at distance 30-35 cm and tetryl (in M-14 mine) at distance 25 cm, which clearly demonstrates the advantages of the two frequency method for detection of explosives and their technological mixtures. In our work we decided to abandon the use of the phase manipulation sequencies, as it does not show any advantages due to the permanently existing inhomogeneity of the field at the locus of a nuclei in the compound of the technological mixture of explosives.

The system shown at Figure 1 is an updated version of the system described in [Grechishkin (1993,1997)], and differes from the latter by its higher phase stability and, unlike other systems, it provides accurate syncronization from oscillators 1 and 2 in the phase sensitive detector.

As the NQR spectra of ^{14}N in TNT have spectral width up to 100 kHz and show high multiplicity and low signal-to-noise ratio, it makes the problem of detection more complicated what acclaims the new approaches to be applied. Often, Fourier transform usually used after preliminary filtering and time averaging of the noisy NQR signals can be complemented by more elaborate techniques of spectral estimation [Grechishkin

(1993,1997), Yung-Ya Lin (1997), Grechishkin (2000)]. One such technique [Grechishkin (2000,2001)] – the Matrix Pencil Method – can handle NQR signals with SNR of 0.7 with close to 100% certainty of detection. To assist in the application of this method we have used the computer code written on the basis of the the MATLAB package [Yung-Ya Lin (1997)].

Inside more than 100 million deployed mines there are all kinds of the technological mixtures of TNT, RDX, tetryl and different organic compounds. It is well known that impurities have an influence on the intensity of NQR signals:

$$n=n_o e^{-NC}$$

where C is concentration of impurity, N – a scaling factor, which is big only for explosives with isomorphous impurities.

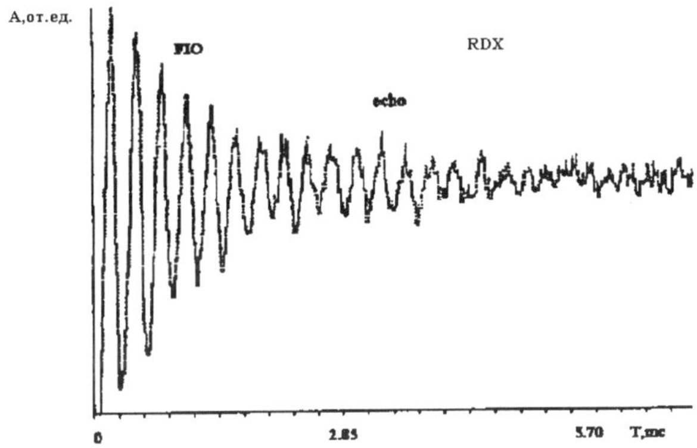

Figure 4. NQR in RDX at 5192 kHz.

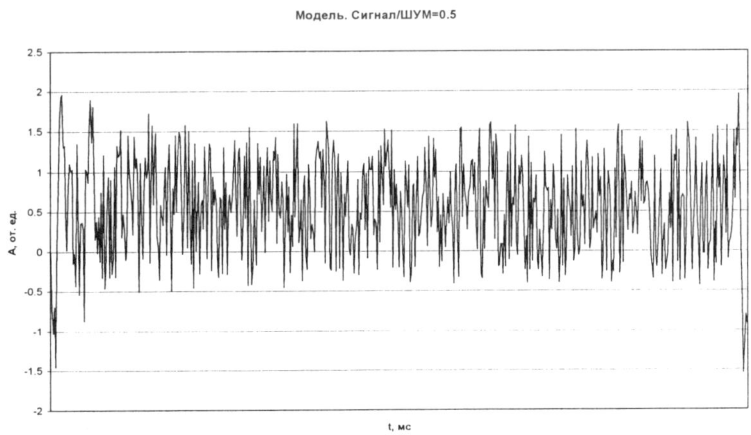

Figure 5. NQR in RDX with SNR = 0.5

Monoclinic TNT presents the most serious difficulties for NQR detection in mines, because of the presence of many organic compounds with such symmetry. We have carried out studies of the influence of different impurities on military TNT. Studies of solid solution of different substances in TNT indicate a change of an integral intensity of NQR due to increase of impurity concentration. In pure p-nitrotoluene T1p is 5 times longer than in 2,4,6-TNT, but T1d in DNQR is the same (0.8 s) at room temperature.

We have used the matrix pencil method (ITMPM) to solve the problem of TNT, and we were able to improve the results for TNT, RDX, PETN and tetryl. For example for remote NQR in RDX at 5192.8 kHz, offset frequency fp=3.5 kHz and line width L=220 Hz we obtaned:

Signal-to-noise ratio (SNR)	ITMPM	LPSVD	FFT
7 dB	fp=3.509 kHz	3.509kHz	3.86 kHz
	L = 304 Hz	326 Hz	700Hz
-3 dB	fp=3.509kHz	2.805kHz	2.41 kHz
	L=1.4 kHz	900 Hz	2.12 kHz

For urotropin the success rate was 89% , if SNR=0.52; for TNT at 100K for SNR=0.8 the success rate was 100%. It is possible to make the commercial NQR mine detector for all mines with TNT, RDX and Tetryl (12,13). The equipment for baggage can also be based on NQR method. If one has NQR mine detector, one can also use it for customs control of the baggage. Figures 4, 5 and 6 show the NQR signals obtained for RDX and urotropin.

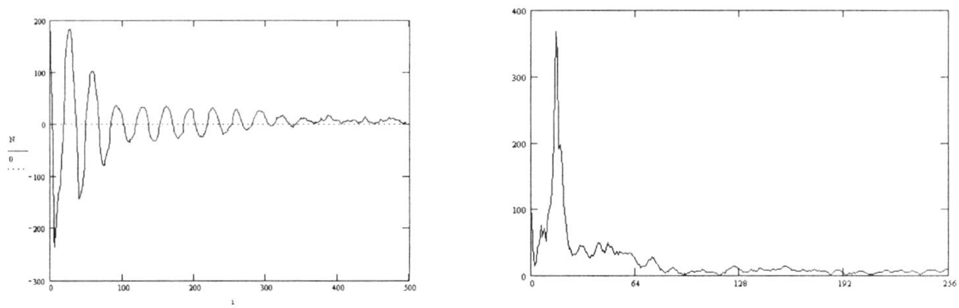

Figure 6. NQR in urotropin at 3306 kHz and ITMPM transformation.

3. Acknowledgements.

The help of my postgraduate students A.Gaevsky, V.Korostelev and O.Selchikhin in carrying our test measurements is gratefully acknowledged. Dr. E.O.Azizov has carried out tests of two-frequency NQR equipment in Cherepovez.

References

Grechishkin V.S., Shishkin E.M., (1973). Org. Magn. Resonance, 5, PP. 1-4.

Azisov E.O., Grechishkin V.S. et al. (1978). Izvest. AN SU, Physics, 42, PP. 2182-2185
Grechishkin V.S., Grechishkina R.V., Izvest. Vuzov, Fizika, N12, 109-110, 1998;
Grechishkin V.S., Z. Naturforsh. Part A: Physics, 45a, N3-4, 559-564, 1990;

Grechishkin V.S., Appl. Phys. A55, 505-507, 1992
Grechishkin V.S., Appl. phys. A58, 63-65, 1994;
Grechishkin V.S., Izv. Vuzov, Fizika, N12, 107-109, 1994.

Anferov V.P., Grechishkin V.S. et al. (1990). Nuclear spin resonance, LGU

Grechishkin V.S., Sinjavsky N.Ja. (1997). Uspehi Fizicheskih Nauk, 189, N10, PP. 195-218; 167, N4, PP. 413-427

Yung-Ya Lin et al, J.Magn.Res. 128, 30-41, 1997.

Grechishkin V.S., (2000) Proc. of the 16th Congress NQR, Leipzig, P. 16, 1999; Proc of the NQR-DLME, Ljubljana, PP.38-40, 2000.

Grechishkin V.S., Grechishkina R.V. (2000,2001). NQR mine-detectors. Patent N2157002, 27.09.2000;
Grechishkin V.S., Korostelev V.V., Device for detection NQR in explosives and narcotics in baggage, patent N2128832, 10.04.1999;
Grechishkin V.S.et al, System NQR for control of baggage and detection of mines. Patent N2165104, 10.04.2001.

DETECTION OF NITROGEN-CONTAINING SUBSTANCES BY MEANS OF A SMALL SIZE ACCELERATOR

V.M. SANIN, A.M.YEGOROV, I.N. ONISHCHENKO, V.JA. MIGALENYA, V.V. ZHUK
Institute of Plasma Electronics and New Methods of Acceleration National Science Center "Kharkov Institute of Physics and Technology Kharkov, 61108, Ukraine

1. Introduction

The problem of explosive detection, including plastic explosive, regardless of an outer shell presence or its composition is urgent at the present time in the context of terrorism and enormous quantity of minefields that are left after wars and local armed conflicts. Explosives detection in the passenger luggage, cargo, and their demining are acute global problems that require elaboration of new detection principles [Gozani (1993)].

We consider the gamma/neutron-location as a possible candidate for detection of explosives/mines, irrespective of their shell composition. The location, as we imply, is the irradiation process of both a soil and explosives/mines with simultaneous detection of reflected or induced gamma-radiation. There are two different location processes: below threshold of nuclear reactions and with the reactions induced by primary irradiation. In the first processes the X-and gamma-ray beams are used to visualize the shape of buried metal and non-metallic subjects. Back-scattered Compton photons are detected if the irradiation is produced by a narrow scanning X-ray beam. If the irradiating photon energy exceeds threshold of the electron-positron pair creation, then photons caused by positron annihilation are detected. This method, similar to a positron tomography, allows to determine the shape of buried subject, too. The methods of shape visualization are considered in section 2.

If the irradiating photon energy exceeds threshold of the nuclear reactions, a nuclear elemental analysis becomes possible. This method based on the (γ,n) reaction is described in section 3. Similar methods using neutron irradiation with the (n,2n) nuclear process are described in section 4.

2. The shape visualization

We believe the shape visualization of the buried subject is a useful informative principle to identify the belonging of a subject to a class sought for [Egorov (1997)]. Visualization methods described in this paper use narrow beams of gamma-rays or X-rays.

H. Schubert and A. Kuznetsov (eds.), Detection of Explosives and Landmines, 227–237.

2.1 THE GAMMA-LOCATION

With this method, an electron accelerator with energy above threshold of the electron-positron pair production is used. The collimated narrow beam from a metal target bombarded with accelerated electrons produces gamma radiation with a continuous spectrum. This beam irradiates the soil containing the searched-for subjects. This leads to the electron-positron pair production by a part of photon spectrum with energy above threshold of pair creation. The annihilating positrons emit two 0.551 MeV gamma-quanta which are detected by scintillation counters synchronously with the irradiation process. The electronic circuit records the coordinates of the irradiating gamma beam and detected photon numbers in the regular intervals simultaneously. This information is used to create the 2D-map of the 0.511 MeV gamma intensity caused by the isotropic annihilation process. The intensity of this process depends on incoming radiation intensity, its energy, and atomic numbers of the soil elements and buried subjects, that allows to map inhomogeneities and hence the shape of hidden subjects. If within the limits of effective depth of irradiation the subject different on its elemental structure from the surrounding ground appears, then the image of this subject lighter or darker (it depends on the elemental structure of the subject) relative to soil background is displayed on the screen. Thus this method is similar to the positron tomography in medicine, but instead of isotopes with positron decay, the process of pair creation by gamma-rays is used.

The space resolution of this method depends on the transverse size of irradiating beam. The processes of interaction of irradiation with ground matter and the object determine the effective depth of vision, the required source power, and the image contrast.

The important factor imaging procedure is the image contrast, which defines the distinguishability of the image details. The required radiation dose and parameters of the setup depend on this value, too. The contrast is given by

$$C = 2(n - n')/(n+n') \qquad (1)$$

where n and n – are the detected number of photons; they correspond to two identical image parts that, in turn, correspond to the resolution element of the image. In order to provide a statistical reliability, the difference $(n - n')$ must exceed the uncertainty σ:

$$(n - n') = m\sigma, \qquad (2)$$

where $\sigma = (n + n')^{1/2}$. At $m = 3$ (3σ criterion) the confidence level is 99.7%, at $m = 2$ the confidence level is equal to 95%. The minimal (threshold) contrast C_{th} may be observed with an average photon number n_a and the $m\sigma$ criterion:

$$n_a = (n + n')/2 = 2(m/C_{th})^2, \qquad (3)$$

In other words, the detected number of photons per resolution element is inversely proportional to the contrast squared. The human eye distinguishes a contrast of ≥ 0.025, but the digital image can be transformed, and any statistically reliable difference becomes visible on the screen. Using the transformed image an operator can conclude whether the subject belongs to looked-for class.

The cross section of the pair production process is proportional to the atomic number, therefore the higher contrast corresponds to a greater difference between the atomic numbers of hidden subject and the soil. In practice, this difference can reach 2.
The simplified scheme of the gamma-locator is shown in Figure 1.

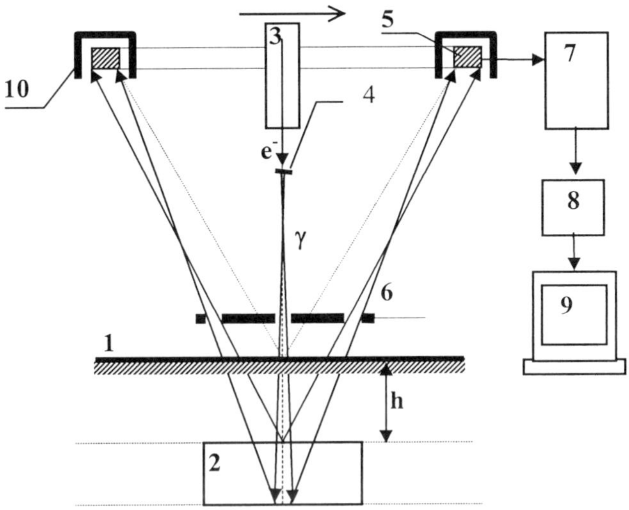

Figure 1. Scheme of a possible gamma-locator version soil; 2- object; 3- electron accelerator; 4- gamma-converter; 5- detector; 6- collimators; 7- electronic circuit 8- AD converter; 9- computer; 10- detector shield; h- depth of vision

Two-dimensional scanning in the horizontal plane can be performed in two ways: (i) through mechanical motion of the entire structure, including the gamma-source, collimators, and detector; (ii) fast scanning of the beam in one of two directions and a slow mechanical scan in the other perpendicular direction. In the last case, not only transverse dimensions of the beam determine the space resolution of the image, but also the beam inclination to the vertical direction. This leads to some decrease a resolution near the end of the scanned area.

The collimator configuration can be made such as to provide a search only within the prescribed depth interval. This can reduce the background from upper soil layers and, hence, increase the image contrast although the angular capture of detector is decreased. Without the limiting collimator (6), the counting rate of detector will be higher but the relative background will also increase.

Although the general process of this method is electron-positron pair production, the Compton back-scattering also occurs. The cross section for this process is small as compared with pair production in this energy range. The energies are approximately twice less than the ones for annihilation quanta. The back-scattered photons are intensely absorbed in a matter and exert only a slight influence on the results. Only some small decrease of space resolution and contrast can occur. The Compton scattering forward almost does not change the energy of photons, and consequently, they can be considered as primary photons.

A version with a fixed gamma-source and a scanning detector is possible, too. In this case, the space resolution is determined by the detector size, and the relative background will be higher, because the irradiated area should be large.

A mobile electron accelerator with energy 5-10 MeV can be used as a gamma-source. It may be an electron linac [Whitham (1991)] or a compact electron microtron [Knapp (1998)]. This method relying on annihilation gamma quanta, is capable to provide high-contrast images of hidden subjects with using moderate power accelerators. Below we present an example of main facility parameters calculated with taking into account the attenuation and formulas 1- 3.

Initial parameters:

Electron energy	6 MeV
Distance between collimator and ground	0.5 m
Radiator thickness	0.01 rad. length
Confidence level	3σ
Soil thickness over the subject	50 gm/cm^2
Average cross-section pair production over energy range 2 – 6 MeV:	
in the object	0.004 cm^2/gm
in soil	0.002 cm^2/gm
Scanning area	1000 cm^2
Angular capture of the detector	0.02 of total sphere

Resulting parameters:

Scanning time	10 s
Contrast	0.6
Space resolution	1.5 cm
Electron current (average)	0.16 mA
Electron beam power	~100 W
Operating ratio of the bremsstrahlung beam	$2\cdot10^{-3}$

2.2 X-RAY LOCATOR

The structure of x-ray locator is similar to the annihilation locator. Only the sources are different, and instead of accelerator, the x-ray tube is used. Calculations indicate the voltage of x-ray tube of 150 kV is preferable. At higher energies of x-rays the attenuation in different materials (in cm^2/gm units) has small differences and materials become weakly distinguishable. In practice, an energy range 0.2- 1.5 MeV is inappropriate for x-ray and gamma location, because attenuation in this range practically depends only on the scattering process, which has the cross sections almost independent of the atomic numbers of the element up to Z = 20 including all main components of the soil and explosive. In the energy band where x-ray energies are lower than 150 kV the photo-absorption process is predominant. This allows identifying impurities in the soil despite the fact that the scattering cross sections are weakly depend on atomic number. The image contrast of this method is lower than that by the annihilation method. The small x-ray energy restricts possibilities of the x-ray location method. The depth up to 15 cm is attainable only. Below we present an example of main facility parameters calculated with taking into account the attenuation.

Initial parameters:

Anode voltage	150 keV
	(energy of the spectrum maximum is 100 keV)
Altitude of the collimator above the ground	15 cm

Distance between the detector and the ground	60 cm
Radius of an annular detector	40 cm
Space resolution	1.5 cm
Diameter of anode focus	5 mm
Scanning area	1000 cm^2
Soil thickness	20 gm/cm^2
Detection angle for back-scattered photons	150°
Confidence level	2σ
Resulting parameters:	
X-ray tube power	2.5 kW
Scanning time	10 s
Image contrast	0.15

This example shows that the gamma-locator provides better parameters, especially the contrast and soil thickness, for visualization of buried subjects than the x-ray locator, but last one utilizes a more simple source for irradiation.

3. Detection of nitrogen

A special feature of all kinds of powerful chemical explosives is the anomalous nitrogen content (from 18% to 43%). There is some correlation between nitrogen, carbon and oxygen. This feature can be useful for identification of explosives. Currently many physical methods of explosive detection are known, including the nuclear elemental analysis using nuclear reactions with gamma-quanta or neutrons [Grodzins (1991,1993)]. Most of the methods need a two-side access to looked-for subjects and are mainly useful for applications at airports [Bach (1993); Vourvopoulos (1993); Trower (1993); Vartsky (1994); Sawa (1993)]. For demining goals only one-side access to the searched subjects is possible.

3.1 NITROGEN DETECTION USING GAMMA-RADIATION

There is a simple method [Habiger (1991)] using an electron accelerator with one-side access. The accelerated electrons strike a heavy metal target, produsing bremsstrahlung with final energy equal to the electron beam energy. The gamma-beam interacts with the explosive nitrogen and excites the photonuclear reaction $^{14}N(\gamma,n)\ ^{13}N$. The stable nitrogen isotope ^{14}N transforms to radioactive isotope ^{13}N, which then decays with a 10 min half-life via positron emission to ^{13}C. The positron immediately annihilates, producing two 511 keV photons that are easily detected using standard scintillation detectors. The threshold of the $^{14}N(\gamma,n)\ ^{13}N$ reaction is about 10.6 MeV. the other elements in explosives and soil have photonuclear reaction thresholds above 13 MeV, as shown in Figure 2, produce no positrons, or have short half-lives. Table 1 lists the reaction thresholds for certain elements and half-lives of corresponding daughter nuclei (β^+ decay).

TABLE 1. Properties of (γ,n) reactions

Element	Threshold, MeV	Daughter nucleus	Half-life (β^+ decay)
^{14}N	10.6	^{13}N	10.1 min
^{16}O	15.7	^{15}O	Stable
^{27}Al	13.1	^{26}Al	6.7 sec
^{28}Si	17.2	^{27}Si	4 sec

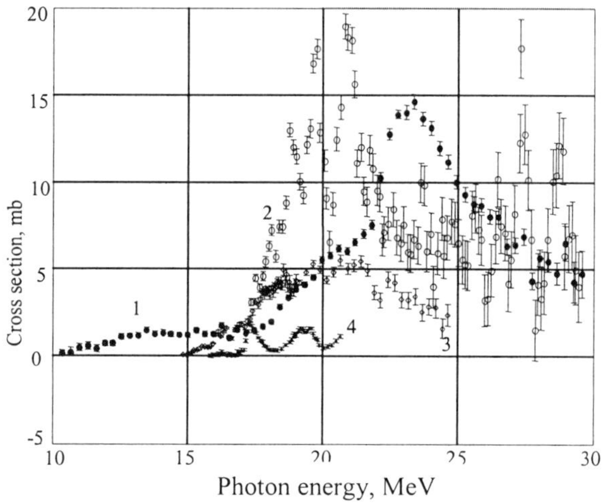

Figure 2. Cross-sections of (γ,n) reactions in ^{14}N - 1, ^{28}Si - 2, ^{27}Al - 3, ^{16}O - 4.

Thus, the electron accelerator with energies of 13.5-14 MeV is optimal to produce the gamma-beam for this purpose. A portable linac [Whitham (1991)] or microtron [Knapp (1998)] can be mounted on a remote-controlled vehicle for demining or placed in airports for inspection of the baggage.

Bremsstrahlung spectrum is very broad and ranging from low energies up to an energy of an electron; but only the energies higher than threshold of the photonuclear reaction $^{14}N(\gamma,n)$ ^{13}N are useful. An electron penetrating into a target dissipates its energy, and eventually, reaches threshold energy. The electrons with energies lower than 10.6 MeV are useless, because the radiation generated by them produces only undesirable background. This, in turn, imposes restriction of the target thickness. This thickness is determined by energy losses in the target. Calculations show that for the tantalum target, the thickness is equal approximately to 0.19 of the radiation length.

A possible version of the demining system similar to the one described in ref. [Habiger (1991)] is shown in Figure 3. Irradiation exposures are made periodically. The detection of the annihilation gamma-quanta from ^{13}N is carried out during the pauses, therefore instantaneous gamma from the processes of electron-positron pair production and the scattered quanta are not detected.

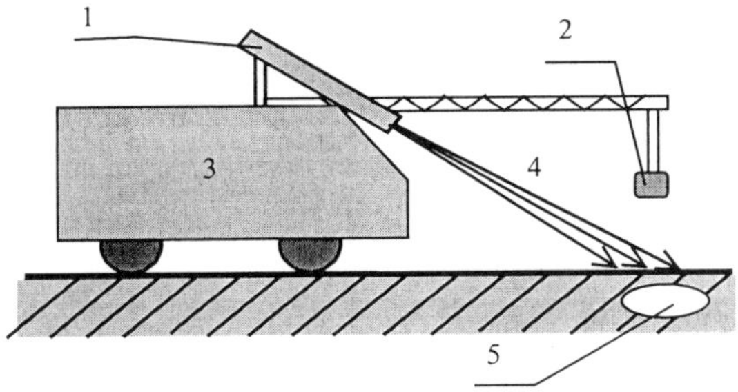

Figure 3. Sketch of the distant demining system; 1- electron linac/microtron; 2- detector; 3- vehicle; 4- bremsstrahlung; 5- mine.

This system with scanning detectors can provide detection of explosives at a distance up to 15 m. In addition to the local metal target we consider the possibility of using the air as a target, too. The required beam power depends on the distance to the mine and the explosive amount, and it can equal 1-15 kW. The spectral distribution was counted with taking into account the electron energy losses. The result is shown in Figure 4.

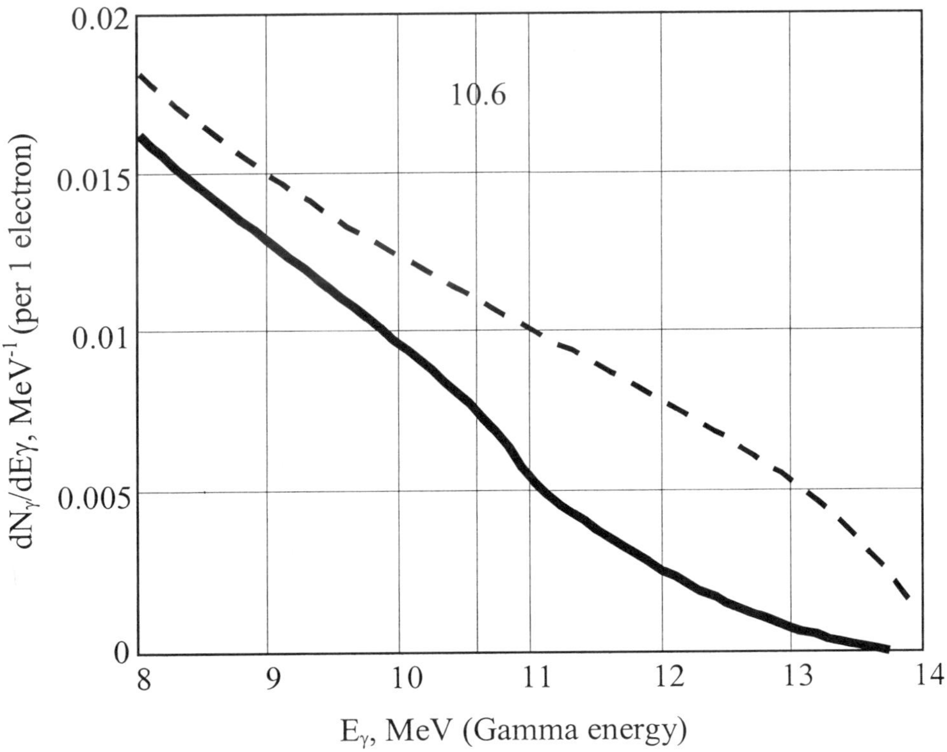

Figure 4. Gamma spectrum: dashed line – without taken into account the electron energy losses (Shiff function); solid line – with taking into account energy losses. Electron energy is 14 MeV.

It can be seen that the energy losses of electrons influence strongly the shape of a spectral distribution and the intensity near the high-energy boundary of the spectrum, which is the working region for demining. The angular distribution of gamma-rays depends strongly on multiple electron scattering in the converting target, and hence, depends on its thickness. In Figure 5, the angular distribution with taken into account the multiple electron scattering but without self-absorption is shown for a tantalum target thickness of 0.19 rad. length (0.78 mm or 1.29 gm/cm^2).

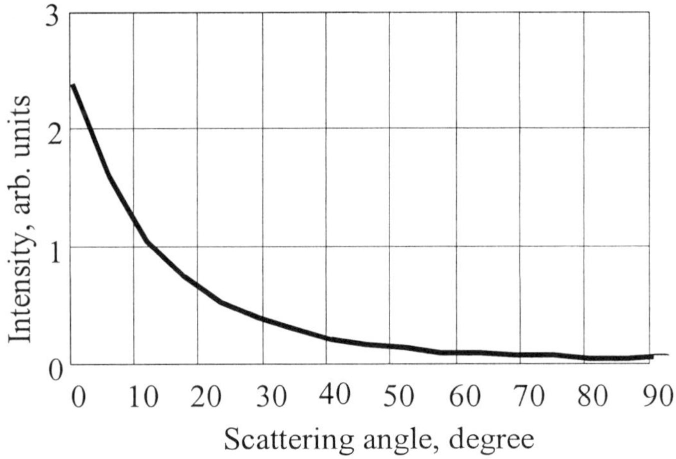

Figure 5. Angular distribution of gamma radiation from the tantalum target.

The figure illustrates good directionality of the gamma beam, which allows illuminate a distant subjects by gamma rays. The ground irradiance distribution caused by a gamma source located at 3 m height above the surface and directed at a distance 15 m from this source is shown in Figure 6.

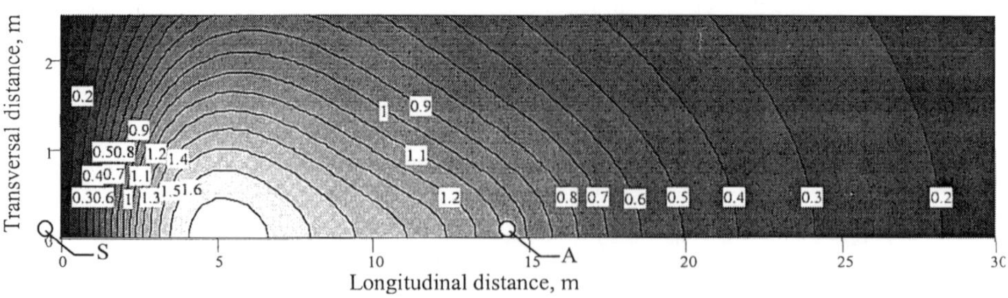

Figure 6 Contour plot of irradiance (half of diagram). The gamma-source is situated 3 meters above a soil surface. S – projection of the source; gamma-beam is directed at the point "A".

The irradiance at this point is accepted as unity. Angular distribution corresponds to the plot shown in Figure 5

The longitudinal profile of this plot is shown in Figure 7.

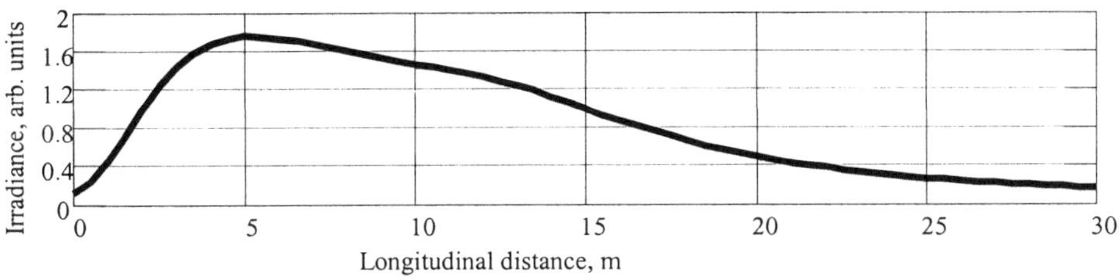

Figure 7 The longitudinal profile of the irradiance

These figures show the irradiance in point of aim is not maximal. It is explained by a slope of beam to the soil surface, the angular distribution of gamma-radiation, and the dependence of irradiance on distance (as $1/R^2$). The distance (R) rapidly decreases because the function $1/R^2$ specifies a rise in the irradiance at small distances.

Figure 6 shows the irradiance does not too much decrease with a transverse distance, therefore a non-collimated gamma beam is preferable. The operating ratio of the gamma beam is in this case higher than when a scanning narrow beam is used. Below we present an example of calculated parameters provided by this method with taking into account the attenuation:

Initial parameters:

Electron energy	14 MeV
Electron current	0.2 mA
Radiator thickness	0.19 rad. length
Detector area	1000 cm^2
Nitrogen mass	20 gm
Distance between explosive and source	15 m
Distance between explosive and detector	0.7 m
Irradiation time	100 s

Result:

Detector's counting rate	13 s^{-1}
Time of explosive detection	~30 s

The scanning detector (NaI) is located at a jib. It has passive and active protective screens. The counting rate may be increased by increasing the electron current, the detector area or by decreasing distance between the explosive and the detector.

Thus, this method is in principle acceptable to detect antipersonnel mines at a distance of 15 m. The detection of explosives at this distance is an acute problem because there are mines that can shoot filaments within this distance.

3.2 NITROGEN DETECTION USING NEUTRON RADIATION

The same atomic nucleus ^{13}N can be produced in the reaction $^{14}N(n,2n)^{13}N$. All the following processes are the same as described in section 3.1. Neutron source, sealed neutron tube for example, is used in this case instead of gamma source based on the electron accelerator. The advantage of this method is that the neutron tube is a simpler instrument as compared with the electron accelerator and has smaller overall dimensions and weight. It can be directly located on a jib near an examined object. The disadvantage is that the neutron source is isotropic and does not allow irradiating a large area of the ground at large distances with an acceptable irradiance. Neutrons with an energy 14.1 MeV from the ^{3}H (d,n) ^{4}He reaction are used to irradiate the soil and the looked-for subject.

In the following calculations we put that the total cross section of the $^{14}N(n, 2n)^{13}N$ reaction is 7 mb. Below we give calculated parameters of a setup using this method.

Initial parameters:

Neutron intensity from the source	10^{13} n/s
Nitrogen mass in explosive subject	20 gm
Detector area	1000 cm^2
Irradiation time	10 s
Height of source above the ground	0.7 m
Height of detector above the subject	0.7 m

Resulting parameters:

Detector's counting rate	88 s^{-1}
Time of explosive detection	1 min (including irradiation time)

The use of BGO scintillation detector by this method is preferable because the neutron source and the detector are close to each other, therefore NaI scintillator will be activated as a result of interaction with neutrons. Increasing of the irradiation time and increasing the source intensity can increase the counting rate. Currently, source with intensity of 10^{14} n/s is under design for boron neutron capture therapy (BNCT) application [Verbeke (1999)].

Thus, this method is applicable for demining objects. Note, one of the disadvantages in this method as compared with gamma-radiation method is that it calls for a greater irradiation time du to a small area under irradiation during one scan.

4. Conclusions

Methods of the visualization of hidden subjects utilizing x-ray and gamma-ray irradiation are proposed. This methods use the x-ray tubes with energy of 150 keV and gamma-ray sources with energies 5-10 MeV above the threshold of the electron-positron pair production, but these energies should be below thresholds of the nuclear reactions at nuclei with ($Z \leq 20$). The x-ray method provides detection of burred subjects under 20 gm/cm^2 soil thickness. The optimal voltage of the x-ray tube is 150 kV. The gamma-ray method can provide the high-contrast visualization of burred subjects under soil thickness up to 50 gm/cm^2, for reasonable time, with using of the electron accelerator of modest power.

Methods of identification of nitrogen-containing subjects utilizing the accelerator-based gamma-ray source and neutron source for irradiation are considered. They are based on the nuclear elemental analysis and utilize (γ,n) and (n,2n) reactions at the nitrogen nucleus respectively. The results of calculations of main parameters of these methods are presented. They show that the accelerator-based gamma-ray source with electron energy 14 MeV can provide the distant identification of small mass explosives within the distances up to 15 m. Neutron intensity of 10^{13} - 10^{14} from sealed neutron tube allows detect the nitrogen-containing explosive for reasonable time, too.

5. Acknowledgment

The authors are thankful to B.A. Nemashkalo for useful discussion of problems relating to this paper.

References

Boch P., Ma J.L., "Chemical Weapons Detection by Fast Neutron Activation", NIM B79 (1993) p.p.605-610.

Egorov A.M., Sanin V.M., "Underground gamma location", Voprosi Atomnoy Nauki i Techniki, serija: Yadero- phizicheskije issledovanija, vipusk 4,5 (31,32), (1997), Kharkov,(in Rus.) p.p.187-189.

Gozani T., "Advances in Accelerator Based Explosives Detection Systems", NIM B79 (1993) p.p.601-604.

Grodzins Lee, "Detecting Contraband in Luggage", NIM B79 (1993) p.p.597-600.

Grodzins Lee, "Nuclear Techniques for Finding Chemical Explosives in Airport Luggage", NIM B56/57 (1991) p.p.829-833.

Habiger K.W., Clifford J.R., Miller R.B., McCullough W.F., "Explosive Detection with Energetic Photons", NIM B56/57 (1991) p.p.834-838.

Knapp E.A., Saunders A.W., Shvedunov V.I., Sobenin N.P., Trower W.P., "A Mobil Racetrack Microtron" NIM B 139 (1998) p.p.517-521.

Sawa Z.P., PFN GASCA Technique for Detection of Explosives and Drugs", NIM B79 (1993) p.p.539-596.

Trower W.P., The Nitrogen Camera and the Detection of Concealed Explosives", NIM B79 (1993) p.p.589-592.

Vartsky D., Engler G., Goldberg M.B., "A Method for Detection of Explosives on Nuclear Resonace Absorption of Gamma Rays in ^{14}N", NIM A348 (1994) p.p.688 – 691.

Verbeke J.M., Lee Y., Leung K.N., Vujic J., Williams M.D., Wu L.K., Zahir N., "Neutron Tube Design Study for Boron Neutron Capture Therapy Application", Proceedigs of the 1999 Particle Accelerator Conference, New York, 1999, pp.2540-2542.

Vourvopoulos G., "A Pulsed Fast-Thermal Neutron System for the Detection of Hidden Explosives", NIM B79 (1993) p.p.585-588.

Whitham K., Miller R.C., Anamkath H., Clifford J.R.,Miller R.B. Habiger K., "Linear Accelerator for Explosive Detection", NIM B56/57 (1991) p.p.825-828.

METHODS OF BUILDING A DATABASE OF RADIOLOCATION CHARACTERISTICS OF EXPLOSIVE SUBSTANCES FOR THEIR LOCALIZATION AND IDENTIFICATION

V. KAPLUN, S. NESTEROV
2nd Central Scientific Research Institute, Ministry of Defense of Russian Federation
170005, Tver, Russia

One of the possible ways of detecting explosive substances is using radars for sub-surface scanning. This allows both establishing the presence of explosive substances or landmines in the certain volume (soil, concrete, brickwork, etc.) and obtaining the radio-image of the inspected area. The image is constructed by using super-high resolution at large distances due to linearly frequency-modulated and super-wide-band signal and high angular resolution due to aperture synthesis technique.

The successful application of radio-imaging systems to explosives detection and imaging is based on reflection and scattering of the incident waves from the explosive material. Hence, the most important task is to create a database of reflective characteristics of various explosives in different surrounding media and probing conditions.

Such characteristics are usually referred to as radiolocation characteristics of objects. To obtain radiolocation characteristics of objects one must use dedicated measurement installations. One of the possibilities is to use the Reference Radiolocation Measurement Complex of the 2nd Central Research Institute of Ministry of Defense of Russian Federation, which allows one to directly measure radiolocation characteristics of explosive substances. The view of the antennae setup and an example of mounting the inspected object are shown at Figure 1.

Figure 1. View of the antennae setup (left) and an example of the mounting of the investigated object (right).

H. Schubert and A. Kuznetsov (eds.), Detection of Explosives and Landmines, 239–241.

Figure 2. The control center of the Reference Radiolocation Measurement Complex.

The Complex has similar characteristics to those of the best foreign analogues, is the only such installation in Russia and CIS states, and is certified and widely used by the Russian Army. In our work we rely on more than 30 years of experience in measurements of radiolocation characteristics of different objects.

All the operations are controlled from the control center (see Figure 2).

The antennae system is calibrated by using a collection of calibration objects (see Figure 3).

Figure 3. A collection of investigated objects (top), and calibration objects (bottom).

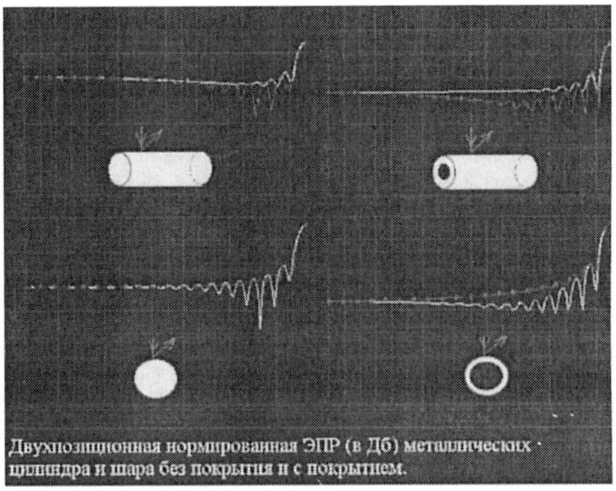

Figure 4. Example of using mathematical methods to model the radiolocation characteristics of different objects.

The work done at the Complex is based on theoretical calculations, which are used to model the characteristics of the investigated objects. Figure 4 shows the capabilities of the mathematical modeling.

Combination of measurements and theoretical analysis of the radiolocation characteristics of explosive substances allows one to get the following information:

- Create a database of radiolocation characteristics of explosives in the whole range of wavelengths – from millimeters to meters.
- Design and build a system for obtaining a radio-image of explosives in different surrounding media. Such systems can be made mobile and be installed on vehicles, helicopters, air-ships, submarines, etc.
- Investigate the possibility of masking the explosives to make its identification by radiolocation technique difficult or impossible.

RADIATION BEAM TECHNOLOGIES IN DETECTION AND DESTRUCTION OF EXPLOSIVES

NIKOLAY A. IVANOV
Department of the Experimental Nuclear Physics, St.Petersburg State Technical University
Politechnicheskaya str., 29, St.Petersburg, 195251, Russia
VALENTINE I. NOVIKOV
Department of the Experimental Nuclear Physics, St.Petersburg State Technical University
Politechnicheskaya str., 29, St.Petersburg, 195251, Russia
ALEXEY A. SHEMAEV
Research Institute of Special Materials
B. Sampsonievski prosp., 28a, St.Petersburg, 194044, Russia

Analysis of known publications [Harding (1993), Lawrence (1994), Vandyshev (1997)] devoted to this problem shows that in most cases modern technical means can not guaranty desirable accuracy and efficiency in detecting or destroying explosives or blast devices (BD). Hence, updating the known and development of new technologies in the field of security is still an urgent problem.

This work presents the analysis of the experimental data obtained while exposing explosives and blast devices to pulsed laser beam radiation [Ivanov (2001)] and fast electrons [Ivanov (1999)].

Potential of using high-energy beams for detecting and destroying explosives and blast devices is based on the fact that beam radiation causes exothermic chemical reaction of explosive molecule decomposition (or synthesis) [Andreev (1960), Baum (1975)] followed by a thermal energy release and various kinds of electromagnetic emission. The above-mentioned reactions obviously lead to quantitative and qualitative variations of the secondary electromagnetic emission radiated by conventional and explosive substances.

In [Ivanov (2001)] a neodymium laser beam with 1.06 μm photon wave length was used for penetrating samples of various explosives and conventional objects. The emitted infrared radiation in the wavelength range of 2μm - 4μm was recorded. It has been found (see Figure 1) that the IR radiation emitted by the explosive samples exposed to a laser pulse with duration 300 μs and energy 1 J was several times higher than the IR radiation exit dose emitted by conventional objects (carton, wood, wax, soap, aluminum, iron) subjected to the same laser pulse. Besides, the time dependence diagram for the explosives secondary radiation signal differs significantly from that for conventional objects, mainly in the duration of the secondary radiation and the existence of the long-time signal decrease component with characteristic duration 1 ms.

The obtained result can be explained by the theory of "hot points" [Andreev (1960), Baum (1975)]. According to this theory, in solid explosives with temperature T_1 the

H. Schubert and A. Kuznetsov (eds.), Detection of Explosives and Landmines, 243–246.

points with increased temperature T_0 occur under the external action. The further temperature changes in such points depend on the dimensionless parameter

$$\delta = QEr^2 z e^{-E/RT_0}/(\lambda R T_0^2),$$

where R – universal gas constant, r – radius of heated region, Q – heat radiation effect per unit volume, E – explosive activation energy, λ – explosive heat conductivity coefficient, z– pre-exponential function.

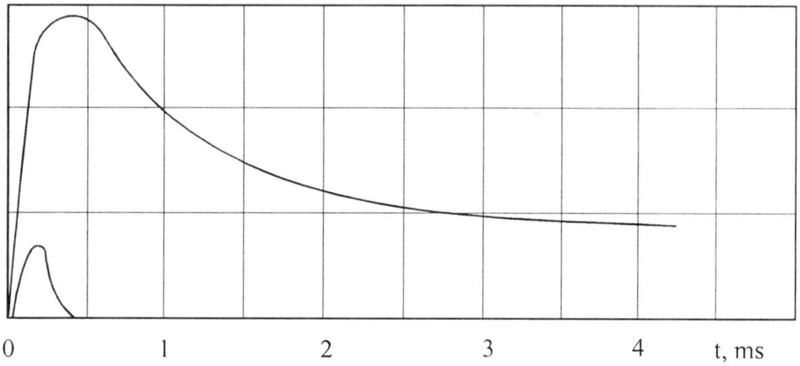

Figure 1. Typical time-dependence of IR radiation dose emitting by the samples exposed by the laser pulse of 300 μs: 1 - conventional objects, 2 - explosives.

At small values of δ, the temperature in the center first starts to grow, and then drops. At high δ exceeding some threshold values δ_{cr} the temperature in the center is growing monotonously up to explosive inflammation, which occurs at critical temperature (T_{cr}) which is for most explosives between 700-900^{0}K (radius of "hot point" is between 10^{-3}-10^{-5} cm).

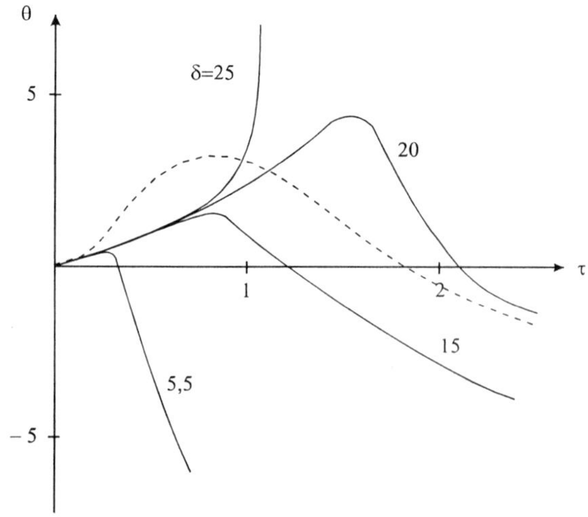

Figure 2. Solid curves - the dependence of non-dimensional temperature in the center of the "hot point" θ of non-dimensional time τ for δ rated between 5 to 25 at $\theta_0 = 10.35$ [Baum (1975)], dotted curve - time dependence of the IR dose emitting by the explosive samples [Ivanov (2001)].

Figure 2 presents dimensionless temperature $\theta=(E/RT_0^2)(T-T_0)$ dependence of dimensionless time $\tau=QEze^{-E/RT_0}/(c\rho RT_0^2)t$ for various δ values obtained from heat balance equation [Baum (1975)], where c, ρ – the explosive specific heat capacity and density. The same figure shows the graphic of the IR radiation emitted by the explosive samples that is close to theoretical θ-dependence of τ at $\delta\approx12$, $T_0\approx500\ ^0K$, $\theta_0=(E/RT_0^2)(T-T_1)\approx19$. Calculation is given for typical explosive parameters.

Explosives also have been exposed to electron beam radiation. The diameter of the beam exceeded cross section of samples. Besides, the electron path inside the samples [Kerim-Markus (1987)] exceeded the sample thickness. As a result, the whole volume of the exposed sample was heated [Andreev (1960), Baum (1975)].

In practice, heat sensitivity of an explosive is characterized by the flash point temperature T_f. A sample is heated up to its flash point and inflames at some given time t_{lat}. At such temperature heat removal is less than heat income. At small values of t_{lat} (~μs) flash point temperature for most explosives is 500-700 ^{0}K.

The speed of heating dT/dt and time of exposing t_{exp} of the samples to the electron beam up to their flash points are found from the following equations:

$$dT/dt = [(dE/dx)I]/(Sc) \qquad (1)$$

$$t_{exp}=T_{cr}/(dT/dt) \qquad (2),$$

where (dE/dx) – ionization loss of electrons; I, S – the flow and the cross section of the beam; c – specific heat capacity.

Calculations done using typical values for explosives (dE/dx) [Kerim-Markus (1987)] and c [Andreev (1960), Baum (1975)] and electron beam characteristics from [Ivanov 1999] have shown that the time of the samples heating in the first set of tests is 10-15^{0}K per second. It is not enough to reach the flash point for most samples during 1 min of exposing. In the second set of tests the time of the samples heating was 25-30^{0}K per second, which should be enough for the samples to reach the critical temperature in less than 20 seconds.

For example, T_{cr} for hexogen is 530^{0}K, for TNT – 750^{0}K " [Andreev (1960), Baum (1975)]. However, the above-mentioned figures do not take into account the effect of heat removal from the surface of the exposed samples. Thus, the time of exposing the samples to the beam up to the flash points in both sets of experiments was longer than calculated from (1), (2), especially for the explosives with high values of T_{cr}, which agrees with the results described in [Stolovy1983].

References

Harding G., editor, (1993) Substance Detection Systems // SPIE Proceedings, v.2092, 550 p., 1993 . Editors: G.Harding, Philips GmbH, Hamburg, Federal Republic of Germany; R.C.Lanza, Massachusetts Institute of Technology, Cambridge, MA, USA; L.J.Myers, Auburn Univ., Auburn, AL, USA.

Lawrence A.H., editor, (1994), Cargo Inspection Technologies // SPIE Proceedings, v.2276, 476 p., 1994. Editors: A.H.Lawrence, Canadian Customs Lab., Ottawa, ON, Canada. Meeting Date: 07/24- 07/29/94, San Diego, CA, USA.

Vandyshev B.A. (1997), Vhodnoj kontrol na vzrivoopasnost. Sistemy bezopasnosti, sviazi, i telekommunikatsij, 1997, # 1(april-may), p.24-27.

Ivanov N.A., Ushakov A.Yu., Shemaev A.A. Kineticheskij, (2001), Metod obnaruzhenija vzrivchatih veschestv // Proc. of the 4th All-Russia Conference "Aktualnije problemi zaschiti I bezopasnosti" , 3 - 6 April, 2001, edited by M.B.Silnikov, St.-Petersburg, 2001, P.198.

Ivanov N.A., Novikov V.I., Sidorov A.V. et al., (1999), Vozdejstvije elektronov na vzrivchatije veschestva. / // Proc. of the 2nd All-Russia Conference "Aktualnije problemi zaschiti I bezopasnosti" 6 – 9 April, 1999, edited by M.B.Silnikov, St.-Petersburg,, 1999, V.2, PP.203-204.

Andreev K.K., Belyaev A.F. (1960), Teorija vzrivchatih veschestv. Moscow, Oborongiz, 1960.

Baum F.A., Orlenko L.P., Stanyukovich K.P., et al., (1975), Fizika vzriva // edites by K.P. Stanukovich, Moscow, Nauka, 1975.

Kerim-Markus I.B., editor, (1987), Tormoznaja sposobnost elektronov I pozitronov. 37 MKRE Report // Edited by I.B. Kerim-Markus, Moscow, Energoatomizdat, 1987.

Stolovy A., Jones E.C., Aviles J.B. et al. (1983), Exothermic reactions in TATB initiated by an electron beam. J. Chem. Phys., 1983, 78 (1), 1 Jan., PP.229-235.

METHODOLOGY AND FIELD EQUIPMENT FOR DETECTION OF EXPLOSIVES, DRUGS AND OTHER SUBSTANCES OF ORGANIC ORIGIN

LERRY MESKHI
Nuclear and Radiation Safety Service of Georgia

Nondestructive detection, localization and control of physical-chemical parameters of substances situated in the vessels with thick walls, including metallic ones, can be performed only by means of methods employing penetrating radiation. It is known that the thicker are the barriers the more penetrating radiation is needed. In the case of one-sided access to the object of control the so-called albedo method, using backscattered radiation is the most promising.

During the last 50 years numerous methods and instruments of this kind have been developed. However limited number of such instruments is in circulation. Among them only few are equipped with neutron sources. It can be easily explained by big volume and weight of a shielding container needed to provide radiation, transportation and operation safety. From this point of view sources of low energy gamma radiation, e.g. Ammericium-241, Barium-133, etc. are more convenient.

Among limited number of autonomous irradiation instruments intended for special search Buster K-910B, developed by CSECO-Campbell Security Equipment Co should be mentioned. This instrument based on gamma albedo method of control is equipped with the source of gamma radiation Barium-133. Because of its technical and operational characteristics and high efficiency it is widely used by custom officers and border-guards in more than 50 states. Owing to the successful construction and radiation safety it has won popularity among customers. However, the physical nature of gamma albedo method itself puts at least two significant limitations on the process of interrogation: small accessible thickness of metallic barriers and impossibility to identify materials hidden behind them, which is caused by the impossibility of identification of reflecting material (e.g. metallic or organic material) on the basis of only the intensity of reflected gamma irradiation.

Using neutron radiation from radioactive sources one can overcome these limitations and broaden both range and effectiveness of radioisotope methods of control, detection and localization in field conditions. Physics of neutron albedo method of measurement differs significantly from gamma albedo method and may provide detection of smaller amount of organic (hydrogen containing) materials as well as their identification behind thicker barriers. It provides a possibility of reliable differentiation between organic materials and metals by the intensity of reflected slow (thermal) neutron flux.

Neutron methods of control are based on the strong dependence of probability of neutron nuclear interactions upon their energy and nuclear characteristics of materials. In particular, identification of organic (substances containing hydrogen) is based on the difference in ability of different organic substances and metallic screens (barriers) to slow down fast neutrons.

H. Schubert and A. Kuznetsov (eds.), Detection of Explosives and Landmines, 247–249.

Over many years different researchers (among them me too) carried out numerous experiments and calculations (Monte-Carlo method) in order to investigate formation of neutron fields in complicated heterogeneous media by different sources of neutrons (which mainly differ by their energetic spectrum). These heterogeneous media imitate different real problems, control conditions, geometry of measurements etc.

The main objective of these investigations and results of analyses can be formulated as detection of local non-uniformities behind metallic barriers by means of albedo neutron methods.

The main conclusion derived from the considered aspects of the albedo neutron method is that it is really possible to identify local non-uniformities caused by the presence of organic materials (substances containing hydrogen) hidden behind thick screens (barriers).

Extra high sensitivity of the method to the presence of such materials behind screens provides their identification not only in the case of their location immediately posterior the screen but also when the distance between the material and the screen makes 100-200mm (depending on the mass of the hidden object and dynamics of the measurement). Solution of such problems in the case of one-sided access to the object is impossible by means of any other physical method of nondestructive control.

Laboratory prototype of neutron albedo equipment was elaborated on the basis of analyses of results of theoretical and experimental investigations. This model was used for optimization of all main technical-operational characteristics of the device to be developed. Technical requirements were set -in accordance with the would be basic application of the instrument such as checking of vehicles e. g. control of chassis, closed cavities of lorries, cars and containers as well as of railway transport, ships, etc.

These main requirements are:

-compactness, lightness, simplicity of operation;

-the area of the working surface should not exceed 100 cm^2;

-the average productivity should be not less than 33 m^2/min;

-the most probable thickness of metallic screens should be 5-15 mm;

-the mass of the identified material hidden behind the 12 mm metallic screen should not be less than 30g;

-the yield of the neutron source should not exceed $5x10^5$ neut/sec.

Examination of the prototype has shown that elaborated method and apparatus not only fully meet above-mentioned requirements but also have a certain capacity for further improvement.

Weight of the instrument with detecting area of 80 cm^2 without the telescopic bar is less than 500g.

The working capacity under given technical parameters is maintained for 50 mm thick metallic and 40 mm thick wooden screens.

Results obtained for Clifornium-252 source with softer neutron energy spectrum were better than results for Plutonium-Beryllium source. Helium-3 neutron counters were used as detectors for backscattered slow neutrons.

A modification of neutron albedo equipment may be used as indicator of liquids and their levels in the reservoirs and tanks with thick walls. It should be mentioned that accuracy of such level indicators lies within ± 1-3 mm and practically does not depend upon physical chemical properties of liquids as well as upon thickness of the walls. The indicator does not need any additional metrology assistance. Several instruments of this kind have been successfully used at the liquid gas storage sites.

Developed neutron albedo apparatus use neutron sources with yields less than (2-5)$\times 10^5$n/s. Distance between operator and source disposed on the telescopic bar makes at least 1 meter. This measure of biological protection by means of distance provides dose limit for population in the case of 6 hours working day.

The above-mentioned certain inconvenience and complicacy of storage and transportation of neutron sources repel potential consumers. The so-called "radio phobia" also should be taken into account. New radioisotope neutron sources can provide real perspective for wide circulation and use of our instrument as well as of other neutron developments in the instrument-making industry. Sources of such kind are called variable fluency neutron sources "on-off" sources, etc. In practice they can be switched on and off mechanically and are radiation safe during storage or transportation. Introduction of such neutron sources to the wide market will initiate and promote "renaissance" of neutron methods of nondestructive control. With the help of such "on-off" neutron sources we are going to elaborate an instrument for identification of plastic mines in dry soils. Such field instrument will be a modification of neutron albedo devices for identification of local non-uniformities.